高 | 等 | 学 | 校 | 计 | 算 | 机 | 专 | 业 | 系 | 列 | 教 | 材

HarmonyOS 应用程序开发与实战
（Java版）

姚信威　主编

清华大学出版社
北京

内 容 简 介

本书系统全面地讲解在鸿蒙操作系统(HarmonyOS)下基于Java的应用程序开发的基础理论知识,通过丰富、详细的案例向读者呈现HarmonyOS应用程序的开发流程。全书共13章。第1章对HarmonyOS的概念、技术特性以及技术架构进行了综合介绍;第2章以一个简单的Hello World工程为例,介绍HarmonyOS应用程序的开发环境、开发工具以及应用的调试过程,并对HarmonyOS的工程结构进行讲解,使读者能更好地切入和理解后续章节的学习内容;第3章详细介绍HarmonyOS应用程序的一大核心——Page Ability,其是完成后续章节学习的基础;第4~6章分别对布局、组件以及对话框进行系统介绍;第7章介绍HarmonyOS应用程序中多媒体的开发过程;第8、9章介绍HarmonyOS应用程序中数据管理和文件管理的部分;第10章介绍HarmonyOS应用程序中后台任务如何通过Service Ability运行;第11~13章分别介绍三个完整的案例(工大通、定点巡检、多媒体播放器),不仅涉及基本的布局、组件、数据管理等基础知识,还涉及了对设备硬件调用等进阶知识,读者可以在这三个案例的基础上进行二次开发,使其功能更加丰富,更具有实用性和应用性。

本书主要面向鸿蒙应用的入门开发人员,也可作为高校教材或培训机构的参考用书。

本书封面贴有清华大学出版社防伪标签,无标签者不得销售。
版权所有,侵权必究。举报:010-62782989,beiqinquan@tup.tsinghua.edu.cn。

图书在版编目(CIP)数据

HarmonyOS应用程序开发与实战:Java版/姚信威主编. —北京:清华大学出版社,2023.6
高等学校计算机专业系列教材
ISBN 978-7-302-63340-2

Ⅰ.①H… Ⅱ.①姚… Ⅲ.①移动终端-应用程序-程序设计-高等学校-教材 Ⅳ.①TN929.53

中国国家版本馆CIP数据核字(2023)第063528号

责任编辑:	龙启铭
封面设计:	何凤霞
责任校对:	韩天竹
责任印制:	丛怀宇

出版发行:清华大学出版社
网　　址:http://www.tup.com.cn,http://www.wqbook.com
地　　址:北京清华大学学研大厦A座　　邮　编:100084
社 总 机:010-83470000　　邮　购:010-62786544
投稿与读者服务:010-62776969,c-service@tup.tsinghua.edu.cn
质量反馈:010-62772015,zhiliang@tup.tsinghua.edu.cn
课件下载:http://www.tup.com.cn,010-83470236

印 装 者:三河市天利华印刷装订有限公司
经　　销:全国新华书店
开　　本:185mm×260mm　　印　张:17.75　　字　数:435千字
版　　次:2023年8月第1版　　印　次:2023年8月第1次印刷
定　　价:59.00元

产品编号:099787-01

前言

HarmonyOS是华为自2012年开发的一款可兼容Android应用程序的跨平台分布式操作系统。自从华为对外流出鸿蒙操作系统(HarmonyOS)相关的设计概念以来,HarmonyOS就引起了广泛关注,它被认为是国产新一代操作系统的希望,是一款"面向未来""面向万物互联"的全场景分布式操作系统。2019年8月,华为在开发者大会上正式发布HarmonyOS,并将该系统开源,供广大开发者学习。2020年9月,华为在开发者大会上发布了HarmonyOS 2.0,推出应用开发者Beta版本,并在同年12月推出了手机开发者Beta版。2021年10月,华为正式发布HarmonyOS 3.0 Beta版,这意味着HarmonyOS整体开发环境和SDK支持也逐步趋于成熟。HarmonyOS的产生体现出"坚持守正创新"的二十大精神,能够培养学生勇于进行理论探索和创新的能力。

HarmonyOS与Android和iOS一样,是独立的操作系统,都支持多种硬件设备,但搭载HarmonyOS的每个设备都不是孤立的,在系统层多终端融为一体,成为"超级终端",终端之间能力、资源可以互助共享。这是HarmonyOS独有的特性,即华为提出的"万物互联"的概念。随着5G时代以及以后的6G时代、7G时代的来临,更多的硬件设备将得到网络支持,而有了HarmonyOS,这些设备不再是独立的个体,而是物联网的一部分,这些物联网设备所产生的数据将在全球物联网大脑中流动,其产生的力量将难以置信。

本书针对HarmonyOS SDK 4版本,对HarmonyOS的应用开发基础进行了梳理和介绍,并搭配对应案例进行讲解,帮助读者快速掌握。本书最后提供了三个完整的HarmonyOS应用程序的开发案例,供读者学习。

全书共13章。第1章对HarmonyOS的概念、技术特性以及技术架构进行了综合介绍,让读者对HarmonyOS有一个全面的了解。第2章以一个简单的Hello World工程为例,介绍HarmonyOS应用程序的开发环境、开发工具以及应用的调试过程,并对HarmonyOS的工程结构进行讲解,使读者能更好地切入和理解后续章节学习内容。第3章详细介绍HarmonyOS应用程序的一大核心——Page Ability,其是完成后续章节学习的基础,使读者更好地学习后续内容。第4~6章分别对布局、组件以及对话框进行系统介绍。第7章介绍HarmonyOS应用程序中多媒体的开发过程。第8、9章介绍HarmonyOS应用程序中的数据管理和文件管理的部分。第10章介绍HarmonyOS应用程序中后台任务如何通过Service Ability运行。第11~13章分别介绍三个完整的案例(工大通、定点巡检、多媒体播放器),不仅涉及基本的布局、组件、数据管理等基础知识,还涉及了对设备硬件调用等进阶开发知识,读者可以在这三个案例的基础上进行二次开发，使其功能更加丰富。

  本书非常适合初学者入门，在内容讲解上，针对代码部分也采用循序渐进的方式进行讲解，保证读者能够根据提供的代码逐步掌握书中的知识点，并且提供直观的可视化效果作为参考。

  在本书编写时，HarmonyOS 的应用开发能力和系统还未完全成熟，处于快速更新迭代的状态，因此本书内容的广度和深度有限，仅涉及 HarmonyOS 应用开发中的一些基础核心功能。另外，在编写过程中，由于 HarmonyOS 的多次更新迭代，本书的代码也经历了多次测试和更改，因此读者在学习过程中也难免会碰到一些问题，还望读者见谅，也欢迎随时联系我们反馈问题。在此，也要感谢李强、邢伟伟、黄胜东、齐楚锋、刘大勇、房立国、王能森、杨望旭、刘旭、何川、张雨辰、章锴杰、林朗、陆琦超、周倩、王鸣飞、袁知恒等在书稿的编写、校对、核验以及本书案例代码的编写、调试等方面做出的贡献。

编 者

2023 年 3 月

目 录

第 1 章　HarmonyOS 简介　/1

1.1　HarmonyOS 的前世今生 ………………………………… 1
　　1.1.1　HarmonyOS 概述 …………………………………… 1
　　1.1.2　HarmonyOS 全场景战略 …………………………… 1
1.2　HarmonyOS 的技术特性 ………………………………… 2
　　1.2.1　硬件互助,资源共享 ………………………………… 3
　　1.2.2　一次开发,多端部署 ………………………………… 5
　　1.2.3　统一操作系统,弹性部署 …………………………… 5
1.3　HarmonyOS 的技术架构 ………………………………… 6
　　1.3.1　内核层 ……………………………………………… 6
　　1.3.2　系统服务层 ………………………………………… 7
　　1.3.3　框架层 ……………………………………………… 7
　　1.3.4　应用层 ……………………………………………… 7

第 2 章　Hello World　/8

2.1　HarmonyOS 应用程序的开发环境 ……………………… 8
2.2　运行 Hello World ………………………………………… 10
　　2.2.1　创建 HelloWorld 项目工程 ………………………… 10
　　2.2.2　启动模拟器 ………………………………………… 13
　　2.2.3　运行 HelloWorld 工程 ……………………………… 15
　　2.2.4　分析 HelloWord 工程 ……………………………… 15
2.3　HarmonyOS 应用程序运行调试 ………………………… 16
　　2.3.1　设置断点 …………………………………………… 16
　　2.3.2　输出日志 …………………………………………… 18
2.4　HarmonyOS 应用程序设备调试 ………………………… 19
　　2.4.1　生成签名文件 ……………………………………… 19
　　2.4.2　签名 HarmonyOS App …………………………… 25
　　2.4.3　无线真机调试 ……………………………………… 25

第 3 章　窗口　/27

3.1　Page Ability 概述 ………………………………………… 27

3.2 Page Ability 的基本用法 ·· 28
   3.2.1 手动创建 Page Ability 类 ·· 28
   3.2.2 在 config.json 文件中注册 Page Ability ·· 28
   3.2.3 创建布局文件 ··· 30
   3.2.4 静态装载布局文件 ·· 31
   3.2.5 显示 Page Ability ·· 31
   3.2.6 销毁 Page Ability ·· 32
3.3 Page Ability 之间的交互 ·· 33
   3.3.1 Intent 的基本概念 ·· 33
   3.3.2 显式使用 Intent ··· 33
   3.3.3 隐式使用 Intent ··· 35
   3.3.4 Page Ability 之间的通信 ·· 37
3.4 Page Ability 的启动类型 ·· 42
3.5 Page Ability 的跨设备迁移 ··· 43
   3.5.1 跨设备迁移前的准备工作 ·· 44
   3.5.2 获取设备列表 ··· 44
   3.5.3 根据设备 ID 调用 Page Ability ··· 49
3.6 AbilitySlice 间导航 ·· 54
   3.6.1 AbilitySlice 的基础用法 ··· 54
   3.6.2 同一 Page 间导航 ·· 55
   3.6.3 不同 Page 间导航 ·· 58
3.7 生命周期 ·· 59
   3.7.1 Page Ability 的生命周期 ·· 59
   3.7.2 AbilitySlice 的生命周期 ··· 60

第 4 章 布局 /62

4.1 Java UI 框架概述 ·· 62
4.2 方向布局 ·· 63
   4.2.1 支持的 XML 属性 ·· 63
   4.2.2 排列方式 ··· 64
   4.2.3 对齐方式 ··· 66
   4.2.4 权重 ·· 68
4.3 依赖布局 ·· 70
   4.3.1 支持的 XML 属性 ·· 70
   4.3.2 排列方式 ··· 71
4.4 表格布局 ·· 76
   4.4.1 支持的 XML 属性 ·· 76
   4.4.2 设置行列数 ·· 76
   4.4.3 设置布局排列方向 ·· 78

4.4.4　设置对齐方式 ·· 78
　4.5　栈布局 ··· 80
　　　4.5.1　支持的 XML 属性 ·· 80
　　　4.5.2　使用默认布局添加组件 ·· 80
　　　4.5.3　使用相对位置添加组件 ·· 81
　4.6　位置布局 ·· 82
　4.7　自适应盒子布局 ·· 84
　　　4.7.1　常用方法 ·· 84
　　　4.7.2　场景示例 ·· 84

第 5 章　UI 组件　/88

　5.1　展示组件 ·· 88
　　　5.1.1　文本组件 ·· 88
　　　5.1.2　图像组件 ·· 91
　　　5.1.3　进度条组件 ··· 94
　　　5.1.4　圆形进度条 ··· 97
　　　5.1.5　时钟组件 ·· 97
　5.2　交互组件 ·· 101
　　　5.2.1　按钮组件 ··· 101
　　　5.2.2　切换按钮组件 ··· 104
　　　5.2.3　文本编辑组件 ··· 105
　　　5.2.4　单选组件 ··· 107
　　　5.2.5　多选组件 ··· 109
　　　5.2.6　开关组件 ··· 111
　5.3　高级组件 ·· 113
　　　5.3.1　列表组件 ··· 113
　　　5.3.2　标签列表组件 ··· 118
　　　5.3.3　滑动选择器组件 ·· 121
　　　5.3.4　日期选择器组件 ·· 124
　　　5.3.5　时间选择器组件 ·· 127
　　　5.3.6　滚动视图组件 ··· 130

第 6 章　对话框　/133

　6.1　普通对话框 ·· 133
　　　6.1.1　显示一个简单的对话框 ·· 133
　　　6.1.2　为对话框添加"关闭"按钮 ··· 134
　　　6.1.3　为对话框添加多个按钮 ·· 135
　　　6.1.4　调整按钮的尺寸 ·· 136
　　　6.1.5　自动关闭对话框 ·· 138

6.2 定制对话框 ………………………………………………………………………… 139
6.3 Toast 信息框 ………………………………………………………………………… 140
6.4 Popup 对话框 ………………………………………………………………………… 142

第 7 章 多媒体 /143

7.1 音频 …………………………………………………………………………………… 143
　　7.1.1 准备本地音频文件 ………………………………………………………… 143
　　7.1.2 播放本地音频文件 ………………………………………………………… 145
　　7.1.3 暂停和继续播放音频 ……………………………………………………… 146
　　7.1.4 停止播放音频 ……………………………………………………………… 146
　　7.1.5 播放在线音频文件 ………………………………………………………… 146
　　7.1.6 播放音频的完整案例 ……………………………………………………… 146
7.2 视频 …………………………………………………………………………………… 148
7.3 相机 …………………………………………………………………………………… 151
　　7.3.1 拍照 API 的使用方式 ……………………………………………………… 151
　　7.3.2 使用相机需要申请的权限 ………………………………………………… 152
　　7.3.3 拍照的完整案例 …………………………………………………………… 152

第 8 章 数据管理 /160

8.1 轻量级数据存储开发 ………………………………………………………………… 160
　　8.1.1 Preferences 类的基本用法 ………………………………………………… 160
　　8.1.2 响应文件写入动作 ………………………………………………………… 162
　　8.1.3 轻量级存储的移动和删除 ………………………………………………… 164
8.2 关系数据库开发 ……………………………………………………………………… 165
　　8.2.1 使用 SQL 语句操作 SQLite 数据库 ……………………………………… 166
　　8.2.2 使用谓词操作 SQLite 数据库 ……………………………………………… 168
　　8.2.3 使用事务 …………………………………………………………………… 171
8.3 对象关系映射数据库 ………………………………………………………………… 174
8.4 分布式文件 …………………………………………………………………………… 177
8.5 分布式数据 …………………………………………………………………………… 180
　　8.5.1 同步数据 …………………………………………………………………… 180
　　8.5.2 用谓词查询分布式数据 …………………………………………………… 183

第 9 章 Data Ability /190

9.1 Data Ability 概述 ……………………………………………………………………… 190
9.2 Data Ability 中的 URI ………………………………………………………………… 190
9.3 创建 Data Ability ……………………………………………………………………… 191
　　9.3.1 手动创建 Data Ability ……………………………………………………… 191
　　9.3.2 自动创建 Data Ability ……………………………………………………… 192

		9.3.3 创建 DataAbilityHelper	193
9.4	Data Ability 访问数据库		193
9.5	Data Ability 访问文件		198
9.6	Data Ability 跨设备访问		201

第 10 章　Service Ability　/206

10.1	Service Ability 概述		206
10.2	Service Ability 的生命周期		206
	10.2.1	启动状态的 Service Ability	206
	10.2.2	连接状态的 Service Ability	207
10.3	Service Ability 的创建		207
	10.3.1	创建 Service Ability	207
	10.3.2	创建前台 Service Ability	209
10.4	Service Ability 的启动与关闭		210
	10.4.1	启动 Service Ability	210
	10.4.2	关闭 Service Ability	211
10.5	Service Ability 的连接		214
	10.5.1	创建接口定义文件	214
	10.5.2	连接 Service Ability	214

第 11 章　工大通　/219

11.1	功能需求分析		219
	11.1.1	每日一报	219
	11.1.2	通行码领取	220
	11.1.3	数据管理	220
	11.1.4	业务流程图	220
	11.1.5	系统构架图	220
11.2	搭建项目框架		221
	11.2.1	创建项目	221
	11.2.2	配置起始页	222
	11.2.3	配置 App 图标和名称	222
11.3	界面设计		223
	11.3.1	登录界面	223
	11.3.2	主界面	224
	11.3.3	每日一报界面	227
	11.3.4	通行码领取界面	229
11.4	功能实现		230

11.4.1 登录功能 ………………………………………………………… 230
11.4.2 表单信息收集 …………………………………………………… 231
11.4.3 数据管理 ………………………………………………………… 233
11.4.4 表单提交 ………………………………………………………… 236
11.4.5 通行码领取 ……………………………………………………… 236

第 12 章 定点巡检 /239

12.1 功能需求分析 ……………………………………………………………… 239
    12.1.1 设备定位及地图位置展示 ……………………………………… 240
    12.1.2 振动提示 ………………………………………………………… 240
    12.1.3 拍照打卡 ………………………………………………………… 240
    12.1.4 历史打卡记录 …………………………………………………… 240
12.2 搭建项目框架 ……………………………………………………………… 240
    12.2.1 项目架构 ………………………………………………………… 240
    12.2.2 权限设置 ………………………………………………………… 240
12.3 页面设计 …………………………………………………………………… 242
    12.3.1 地图界面 ………………………………………………………… 242
    12.3.2 打卡拍照界面 …………………………………………………… 242
    12.3.3 打卡记录界面 …………………………………………………… 242
12.4 功能实现 …………………………………………………………………… 244
    12.4.1 数据管理 ………………………………………………………… 244
    12.4.2 定位及地图展示 ………………………………………………… 247
    12.4.3 振动器调用 ……………………………………………………… 250
    12.4.4 相机调用 ………………………………………………………… 250
    12.4.5 打卡操作 ………………………………………………………… 252
    12.4.6 打卡记录查询 …………………………………………………… 253

第 13 章 多媒体播放器 /255

13.1 功能需求分析 ……………………………………………………………… 255
    13.1.1 获取读取本地媒体文件权限 …………………………………… 255
    13.1.2 获取本地媒体文件 ……………………………………………… 256
    13.1.3 播放音频和视频 ………………………………………………… 256
13.2 搭建项目框架 ……………………………………………………………… 256
    13.2.1 项目结构 ………………………………………………………… 256
    13.2.2 添加应用权限 …………………………………………………… 256
    13.2.3 配置相关的 abilities ……………………………………………… 257
13.3 界面设计 …………………………………………………………………… 258

13.3.1　欢迎界面 …………………………………………………………… 258
　　13.3.2　媒体列表页 ………………………………………………………… 258
　　13.3.3　播放器界面 ………………………………………………………… 261
13.4　功能实现 ………………………………………………………………………… 263
　　13.4.1　欢迎界面及权限授予 ……………………………………………… 263
　　13.4.2　媒体列表及获取本地媒体文件 …………………………………… 264
　　13.4.3　封装一个Player播放器类 ………………………………………… 268
　　13.4.4　实现PlayerAbility ………………………………………………… 270

第1章

HarmonyOS 简介

为面向万物互联时代,多种终端设备能多场景应用,在传统的单设备系统能力基础上,增加了分布式理念,一款全新的操作系统——鸿蒙操作系统(HarmonyOS)应运而生。

通过阅读本章,读者可以掌握:

➢ 什么是 HarmonyOS。
➢ HarmonyOS 全场景战略。
➢ HarmonyOS 的技术特性。
➢ HarmonyOS 的技术架构。

1.1 HarmonyOS 的前世今生

1.1.1 HarmonyOS 概述

HarmonyOS 是一款面向未来的(万物互联时代)、全新的分布式操作系统。在传统的单设备系统能力基础上,HarmonyOS 创造性地提出了基于同一套系统能力、适配多种终端形态的分布式理念,能够支持手机、平板、智能穿戴、智慧屏、车机等多种终端设备,提供全场景(移动办公、运动健康、社交通信、媒体娱乐等)业务能力。针对不同的用户群体,HarmonyOS 具有以下不同的特性:

(1) 对消费者而言,HarmonyOS 能够将生活场景中的各类终端进行能力整合,形成一个"超级虚拟终端",实现不同终端设备之间的快速连接、能力互助、资源共享,匹配合适的设备、提供流畅的全场景体验;

(2) 对应用开发者而言,HarmonyOS 采用了多种分布式技术,使应用开发与不同终端设备的形态差异无关,降低开发难度和成本,从而让开发者能够聚焦上层业务逻辑,更加便捷、高效地开发应用。不仅如此,HarmonyOS 还提供了支持多种开发语言的 API,供开发者进行应用开发。支持的开发语言包括 Java、XML(Extensible Markup Language)、C/C++、JS(JavaScript)、CSS(Cascading Style Sheets)和 HML(HarmonyOS Markup Language);

(3) 对设备开发者而言,HarmonyOS 采用了组件化的设计方案,可根据设备的资源能力和业务特征灵活裁剪,满足不同形态终端设备对操作系统的要求。一套操作系统可以满足不同能力的设备需求,实现统一操作系统,弹性部署。

1.1.2 HarmonyOS 全场景战略

为了打造未来 5G 全场景智慧生活,华为提出了"1+8+N"产品战略,打造以个人或家庭为中心的生活全场景，如图 1-1 所示。其中，"1"代表智能手机，"8"代表平板、PC、耳机、车机、手表、眼镜、音响、智慧屏，"N"在"8"的基础上，又进一步连接更多的设备，主要包括以下 5 个模块：

图 1-1　全场景战略示意图

（1）移动办公：连接投影仪、打印机之类的设备；
（2）智能家居：连接摄像头、扫地机之类的设备；
（3）运动健康：连接智能秤、血压计之类的设备；
（4）影音娱乐：连接视频、游戏之类的设备；
（5）智能出行：连接车辆信息、地图之类的设备。

"1"和"8"是华为自己构建的，以手机的优势向外围扩展，而"N"则是所有生态系统合作伙伴提供的智能设备，基于用户为中心的家庭场景，提供全场景的视听、娱乐、社交、教育和健康等解决方案，从而更好地迎合时代更新换代的消费升级。

在"1＋8＋N"全场景战略中有一个非常重要的应用就是 HUAWEI SHARE。HUAWEI SHARE 是基于华为自有的"1"和"8"的产品，能够实现一碰传输文件，一碰传输音频，一碰连接网络，多屏协同等创新的功能。HUAWEI HiLink 可以将"1＋8"的终端与海量的终端设备"N"连接，能够实现设备一键操控、语音交互、场景联动等极致体验。

1.2　HarmonyOS 的技术特性

相对于市面上已有的操作系统，尤其是相对于安卓系统，HarmonyOS 具备以下 3 个主要特征：
（1）以分布式为基础的硬件互助，资源共享；
（2）应用一次开发，多端部署；
（3）系统与硬件解耦，统一操作系统，弹性部署。

1.2.1 硬件互助，资源共享

多种设备之间能够实现硬件互助、资源共享，依赖的关键技术包括分布式软总线、分布式设备虚拟化、分布式数据管理、分布式任务调度等。

1. 分布式软总线

分布式软总线技术是基于华为多年的通信技术积累，参考计算机硬件总线，在"1＋8＋N"设备之间搭建一条"无形"的总线，具备自发现、自组网、高带宽、低时延、高可靠等特点，如图 1-2 所示。

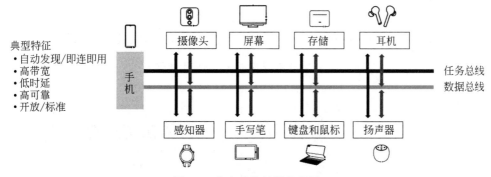

图 1-2　分布式软总线示意图

设备通信方式多种多样（USB、Wi-Fi、BT 等），不同通信方式采用不同的通信协议，使用差异很大且烦琐，同时通信链路的融合共享和冲突无法处理，通信安全问题也不好保证。分布式软总线致力于实现近场设备间统一的分布式通信能力管理，提供不区分链路的设备发现和传输接口。目前实现能力包含：①服务发布，服务发布后周边的设备可以发现并使用服务；②数据传输，根据服务的名称和设备 ID 建立一个会话，这样就可以实现服务间的传输功能；③安全，提供通信数据的加密能力。

分布式软总线是手机、平板、智能穿戴、智慧屏、车机等分布式设备的通信基座，为设备之间的互联互通提供了统一的分布式通信能力，为设备之间的无感发现和零等待传输创造了条件。开发者只需聚焦于业务逻辑的实现，无需关注组网方式与底层协议。分布式软总线的架构示意图见图 1-3。

2. 分布式设备虚拟化

分布式设备虚拟化平台可以实现不同设备的资源融合、设备管理、数据处理，多种设备共同形成一个超级虚拟终端。针对不同类型的任务，为用户匹配并选择能力合适的执行硬件，让业务连续地在不同设备间流转，充分发挥不同设备的能力优势，如显示能力、摄像能力、音频能力、交互能力以及传感器能力等。分布式设备虚拟化示意图见图 1-4。

3. 分布式数据管理

分布式数据管理基于分布式软总线的能力，实现应用程序数据和用户数据的分布式管理。用户数据不再与单一物理设备绑定，业务逻辑与数据存储分离，跨设备的数据处理如同本地数据处理一样方便快捷，让开发者能够轻松实现全场景、多设备下的数据存储、共享和访问，为打造一致、流畅的用户体验创造了基础条件。分布式数据管理示意图见图 1-5。

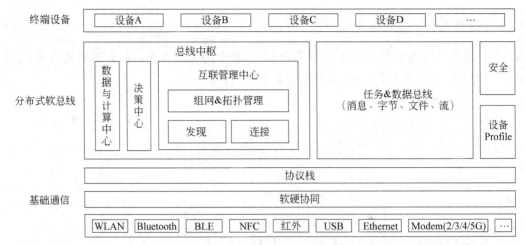

图 1-3 分布式软总线的架构示意图

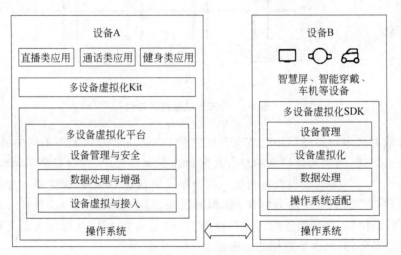

图 1-4 分布式设备虚拟化示意图

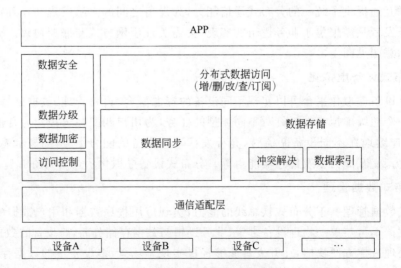

图 1-5 分布式数据管理示意图

4. 分布式任务调度

分布式任务调度基于分布式软总线、分布式数据管理、分布式 Profile 等技术特性，构建统一的分布式服务管理（发现、同步、注册、调用）机制，支持对跨设备的应用进行远程启动、远程调用、远程连接以及迁移等操作，能够根据不同设备的能力、位置、业务运行状态、资源使用情况，以及用户的习惯和意图，选择合适的设备运行分布式任务。

图 1-6 以应用迁移为例，简要地展示了分布式任务调度能力。

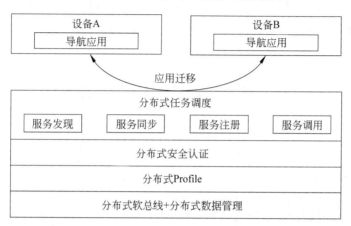

图 1-6　分布式任务调度示意图

1.2.2　一次开发，多端部署

HarmonyOS 提供了用户程序框架、Ability 框架以及 UI 框架，支持应用开发过程中多终端的业务逻辑和界面逻辑进行复用，能够实现应用的一次开发、多端部署，提升了跨设备应用的开发效率。一次开发、多端部署示意图见图 1-7。

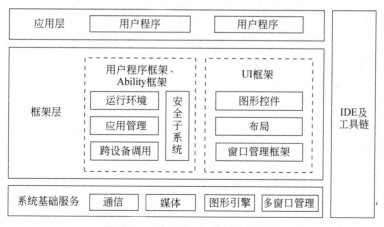

图 1-7　一次开发、多端部署示意图

UI 框架支持使用 Java、JS、TS 语言进行开发，并提供了丰富的多态控件，可以在手机、平板、智能穿戴、智慧屏、车机上显示不同的 UI 效果。采用业界主流设计方式，提供多种响应式布局方案，支持栅格化布局，满足不同屏幕的界面适配能力，在跨设备之间实现共享生态。

1.2.3 统一操作系统，弹性部署

HarmonyOS 通过组件化和小型化等设计方法，支持多种终端设备按需弹性部署，能够适配不同类别的硬件资源和功能需求。支持通过编译链关系去自动生成组件化的依赖关系，形成组件树依赖图，支持产品系统的便捷开发，降低硬件设备的开发门槛。

HarmonyOS 设计上支持根据硬件的形态和需求，选择所需的组件；支持根据硬件的资源情况和功能需求，选择配置组件中的功能集。例如，选择配置图形框架组件中的部分控件；支持根据编译链关系，可以自动生成组件化的依赖关系。例如，选择图形框架组件，将会自动选择依赖的图形引擎组件等。

1.3 HarmonyOS 的技术架构

HarmonyOS 整体遵从分层设计，如图 1-8 所示，从下向上依次为内核层、系统服务层、框架层和应用层。系统功能按照"系统→子系统→功能/模块"逐级展开，在多设备部署场景下，支持根据实际需求裁剪某些非必要的子系统或功能/模块。

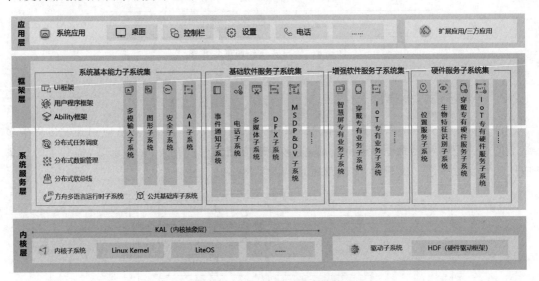

图 1-8 HarmonyOS 技术架构

1.3.1 内核层

HarmonyOS 的内核层由内核子系统和驱动子系统两个子系统构成。

（1）内核子系统：HarmonyOS 的内核不是传统意义上的单内核架构，如 Android 或者 Linux，从图 1-8 可以看出，它采用多内核设计，支持针对不同资源受限设备选用适合的 OS 内核，所以 HarmonyOS 可以同时支持无内存管理单元（MMU, Memory Management Unit）的架构和有 MMU 的架构。内核抽象层（KAL, Kernel Abstract Layer）通过屏蔽多内核差异，对上层提供基础的内核能力，包括进程/线程管理、内存管理、文件系统、网络管理和外设管理等。

(2) 驱动子系统：HarmonyOS 的硬件驱动框架（HDF）是 HarmonyOS 硬件生态开放的基础，提供统一外设访问能力和驱动开发、管理框架。HDF 驱动加载包括按需加载和按序加载两种方式。①按需加载：HDF 框架支持驱动在系统启动过程中默认加载，或者在系统启动之后动态加载。②按序加载：HDF 框架支持驱动在系统启动的过程中按照驱动的优先级进行加载。HDF 框架可以集中管理驱动服务，使用者可直接通过 HDF 框架对外提供的能力接口获取驱动相关的服务。同时 HDF 框架提供统一的驱动消息机制，支持用户态应用向内核态驱动发送消息，也支持内核态驱动向用户态应用发送消息。

1.3.2 系统服务层

系统服务层是 HarmonyOS 的核心能力集合，通过框架层对应用程序提供服务，该层包含以下几个部分：

（1）系统基本能力子系统集：为分布式应用在 HarmonyOS 多设备上的运行、调度、迁移等操作提供了基础能力，由分布式软总线、分布式数据管理、分布式任务调度、方舟多语言运行时、公共基础库、多模输入、图形、安全、AI 等子系统组成。其中，方舟多语言运行时提供了 C、C++、JS 多语言运行时和基础的系统类库，也为使用方舟编译器静态化的 Java 程序（即应用程序或框架层中使用 Java 语言开发的部分）提供运行时；

（2）基础软件服务子系统集：为 HarmonyOS 提供公共的、通用的软件服务，由事件通知、电话、多媒体、DFX（Design For X）、MSDP&DV 等子系统组成；

（3）增强软件服务子系统集：为 HarmonyOS 提供针对不同设备的、差异化的能力增强型软件服务，由智慧屏专有业务、穿戴专有业务、IoT 专有业务等子系统组成；

（4）硬件服务子系统集：为 HarmonyOS 提供硬件服务，由位置服务、生物特征识别、穿戴专有硬件服务、IoT 专有硬件服务等子系统组成。

根据不同设备形态的部署环境，基础软件服务子系统集、增强软件服务子系统集、硬件服务子系统集内部可以按子系统粒度裁剪，每个子系统内部又可以按功能粒度裁剪。

1.3.3 框架层

框架层为 HarmonyOS 应用开发提供了 Java、C、C++、JS、TS 等多语言的用户程序框架和 Ability 框架，两种 UI 框架（包括适用于 Java 语言的 Java UI 框架，适用于 JS、TS 语言的方舟开发框架），以及各种软硬件服务对外开放的多语言框架 API。同时为采用 HarmonyOS 的设备提供了 C、C++、JS 等多语言的框架 API，不同设备支持的 API 与系统的组件化裁剪程度相关。

1.3.4 应用层

应用层包括系统应用和第三方非系统应用。HarmonyOS 的应用由一个或多个 FA(Feature Ability)或 PA(Particle Ability)组成。其中，FA 有 UI 界面，提供与用户交互的能力，而 PA 无 UI 界面，提供后台运行任务的能力以及统一的数据访问抽象。FA 在进行用户交互时所需的后台数据访问也需要由对应的 PA 提供支撑。基于 FA/PA 开发的应用，能够实现特定的业务功能，支持跨设备调度与分发，为用户提供一致、高效的应用体验。

第2章 Hello World

本章介绍如何实现在 HarmonyOS 中运行 Hello World 工程(工程本质上就是一个程序),然后进行调试,最后在模拟机上以及在真机上运行。

通过阅读本章,读者可以掌握:

- 配置 HarmonyOS 的开发环境。
- 创建 Hello World 工程。
- 运行 Hello World 工程。
- 调试程序。
- 真机设备调试。

2.1 HarmonyOS 应用程序的开发环境

HUAWEI DevEco Studio 是基于 IntelliJ IDEA 社区版二次开发的,面向华为终端开发多设备的集成开发环境(IDE),为开发者提供工程模板创建、开发、编译、调试、发布等 E2E 的 HarmonyOS 应用开发服务。由于 HUAWEI DevEco Studio 是基于 IntelliJ IDEA 社区版二次开发的,在使用时会与 IntelliJ IDEA 有很多相似之处,可以让开发人员更快上手开发 HarmonyOS 应用。

DevEco Studio 支持 Windows 和 macOS 系统。读者可以通过下述链接根据需要下载 DevEco Studio 的最新版(https://developer.harmonyos.com/cn/develop/deveco-studio#download)。

打开如上链接,找到如图 2-1 所示的下载列表。单击右侧的下载按钮,即可下载。需要注意的是下载 DevEco Studio 需要登录华为账号,该华为账号还需实名认证,没有账号的可以提前注册一个。

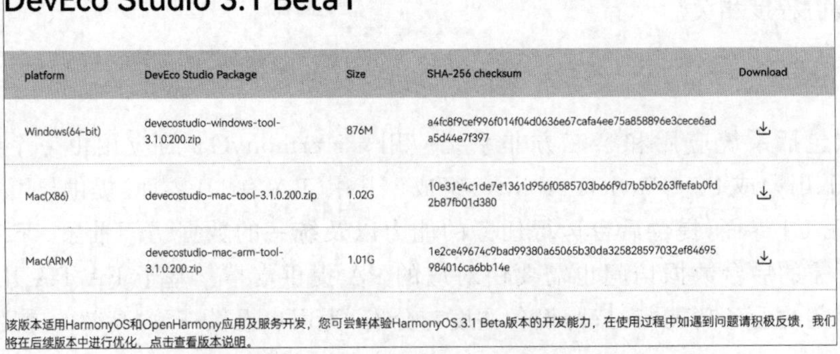
DevEco Studio 3.1 Beta1

platform	DevEco Studio Package	Size	SHA-256 checksum	Download
Windows(64-bit)	devecostudio-windows-tool-3.1.0.200.zip	876M	a4fc8f9cef996f014f04d0636e67cafa4ee75a858896e3cece6ad a5d44e7f397	
Mac(X86)	devecostudio-mac-tool-3.1.0.200.zip	1.02G	10e31e4c1de7e1361d956f0585703b66f9d7b5bb263ffefab0fd 2b87fb01d380	
Mac(ARM)	devecostudio-mac-arm-tool-3.1.0.200.zip	1.01G	1e2ce49674c9bad99380a65065b30da325828597032ef84695 984016ca6bb14e	

该版本适用HarmonyOS和OpenHarmony应用及服务开发,您可尝鲜体验HarmonyOS 3.1 Beta版本的开发能力,在使用过程中如遇到问题请积极反馈,我们将在后续版本中进行优化,点击查看版本说明。

图 2-1 DevEco Studio 下载页面

第 2 章　Hello World

下载的 DevEco Studio 是一个压缩包，需要将其解压。如果使用 Windows 版本，打开的是一个 EXE 安装文件，直接双击安装即可；如果使用的是 macOS 版本，里面是一个 DMG 文件，也是直接双击安装。安装完成后，Windows 版本会在桌面出现一个 DevEco Studio 图标，macOS 版本在应用程序里出现，直接双击图标即可运行。

下面以 Windows 10 版本为例，简单介绍安装步骤。

安装界面如图 2-2 所示，根据操作系统类型选择是否勾选 1，若为 64 位操作系统则勾选 1，否则不勾选。若不清楚系统类型，可通过右击"此电脑"，依次选择"属性"→"关于"查看。再勾选 2，便于安装时自动添加环境变量。若勾选 3，可通过在项目目录处右击打开项目，而不需要通过 DevEco Studio 打开项目，可根据个人使用习惯来勾选。

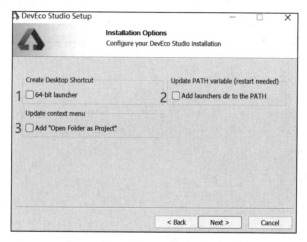

图 2-2　DevEco Studio 的安装界面

运行 DevEco Studio 后，会出现如图 2-3 所示的欢迎界面，如果是第一次使用，那么右侧的 Recent Project 列表为空，显示"No data"；如果使用过 DevEco Studio，那么在右侧就会出现曾经创建过的 HarmonyOS 工程，单击其中一个 HarmonyOS 工程，就会打开该工程。

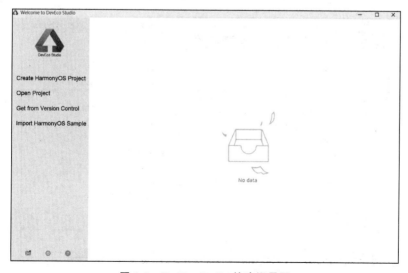

图 2-3　DevEco Studio 的欢迎界面

单击左侧的 Create HarmonyOS Project 创建一个全新的 HarmonyOS 工程，也可以单击 Open Project 打开一个该计算机上所存储过的 HarmonyOS 工程。2.2 节会详细介绍如何使用 DevEco Studio 来开发第一个 HarmonyOS App。

2.2 运行 Hello World

2.2.1 创建 HelloWorld 项目工程

在 DevEco Studio 的欢迎界面（见图 2-4）中单击 Create HamonyOS Project，创建一个全新的 HarmonyOS 工程。也可以在主界面菜单栏选择如图 2-5 所示的 File 菜单，依次选择"New"→"New Project"选项打开 Create HarmonyOS Project 窗口。

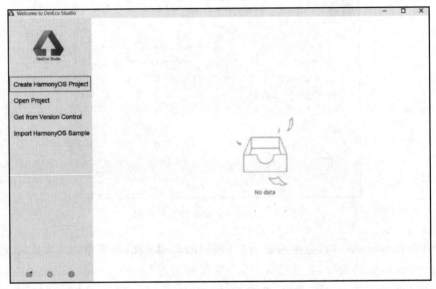

图 2-4　DevEco Studio 的创建界面

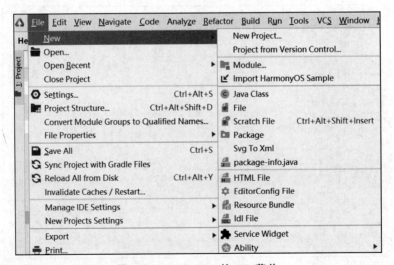

图 2-5　New Project 的 File 菜单

打开的 Create HarmonyOS Project 窗口如图 2-6 所示。

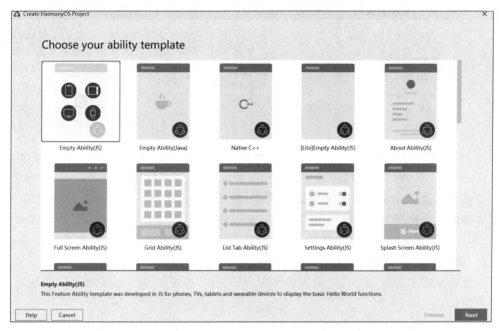

图 2-6　Create HarmonyOS Project 窗口

DevEco Studio 为开发者提供了适用于不同开发需求的模板，详细介绍如表 2-1 所示。

表 2-1　模板介绍

模板名称	支持的设备	支持的开发语言	模板说明
Empty Ability	Phone、Tablet、TV、Wearable	JavaScript	支持低代码开发，用于 Phone、TV、Tablet、Wearable 设备的 Feature Ability 模板，展示了基础的 Hello World 功能
	Phone、Tablet、TV、Wearable、Car	Java	用于 Phone、TV、Tablet、Wearable、Car 设备的 Feature Ability 模板，展示了基础的 Hello World 功能
	Phone、Tablet、Car	eTS	用于 Phone、TV、Tablet 设备的 Feature Ability 模板，展示了基础的 Hello World 功能
Native C++	Phone、Car	C++	用于 Phone、Car 设备的 Feature Ability 模板，作为 HarmonyOS 应用/服务调用 C++ 代码的示例工程，界面显示"Hello from JNI C++ codes"
About Ability	Phone	Java	用于 Phone 设备的 Feature Ability 模板。展示了一个应用的"关于"页模板，提供了应用关于信息的实现方式
	Phone	eTS	
	Phone、Tablet、TV、Wearable	JavaScript	用于 Phone、Tablet、TV、Wearable、Phone 设备的 Feature Ability 模板。展示了一个应用的"关于"页模板，提供了应用关于信息的实现方式

续表

模板名称	支持的设备	支持的开发语言	模板说明
Grid Ability	Phone	Java	用于 Phone 设备的 Feature Ability 模板,使用 XML 来写布局,显示内容为两部分网格表,网格每行显示 4 个项目,网格内元素可进行拖动排序
	Phone	eTS	
	Phone、Tablet、TV	JavaScript	用于 Phone、Tablet、TV 设备的 Feature Ability 模板。展示了一个网格页模板,用于网格应用和状态栏的展示

HarmonyOS App 开发的方式与 Android 开发类似,很多框架也与 Android 开发类似,这主要是为了让 Android 的开发人员更容易上手开发 HarmonyOS App。JavaScript 开发的方式类似于小程序的开发技术,采用的是 Web 栈技术,这主要是针对微信小程序开发人员和 Web 开发人员的。这就意味着,只要是安卓开发人员、微信小程序开发人员或 Web 开发人员,就非常容易上手开发 HarmonyOS App。

根据开发需求选择一个合适的模块作为基础工程,并对其进行配置,如图 2-7 所示是工程配置窗口。

图 2-7 工程配置窗口

配置 HarmonyOS 工程需要指定以下 6 个信息。

(1) Project Name:工程的名称。

(2) Project Type:工程的类型,标识该工程是一个原子化服务(Atomic Service)或传统方式的需要安装的应用(Application)。

(3) Package Name:HarmonyOS 工程的包名,与 Android 中包的含义相同,用于在 HarmonyOS 设备中唯一标识 HarmonyOS App。

(4) Save Location：指定 HarmonyOS 工程保存的目录。注意，这个目录应该是空的，DevEco Studio 会直接将与工程相关的目录和文件都放到这个保存目录中。

(5) Compatible API Version：HarmonyOS SDK 的 API 版本，大多数设备都有 4 和 5 两个版本，可以选择其中一个。

(6) Device Type：选择应用的设备，可以选择 Phone（智能手机），Tablet（平板电脑），TV（电视机，即华为智慧屏以及任何安装了 HarmonyOS 的智能电视），Wearable（智能手表），Car（车载电脑）。

2.2.2 启动模拟器

在没有真机的情况下，只能在 DevEco Studio 模拟器中运行 HarmonyOS App。不过，因为目前 HarmonyOS 没有 x86 版的模拟器，只能通过 ARM 服务器模拟 HarmonyOS 的各种硬件设备。使用 DevEco Studio 中的模拟器需要通过华为账号申请，申请成功后，会在 IDE 中显示一个预览界面。

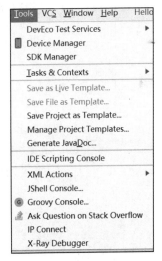

图 2-8　Tools 菜单项

申请的过程：首先在 DevEco Studio 中选择"Tools"→"Device Manager"菜单项，如图 2-8 所示，如果是第一次使用，会出现登录页面，如图 2-9 所示，单击"Login"按钮，跳转至华为账号登录页面，如图 2-10 所示。

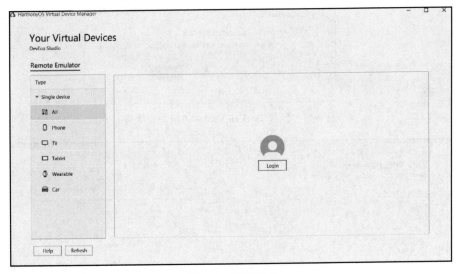

图 2-9　登录页面

如果没有华为账号，可以单击"注册"链接进行注册；如果有华为账号，直接登录即可。成功登录后，会弹出如图 2-11 所示的对话框，提示"HUAWEI DevEco Studio 想要访问您的华为账号"，单击"允许"按钮。

授权成功后，回到 DevEco Studio，会弹出"Virtual Device Manager"对话框，如图 2-12 所示。

图 2-10　华为账号登录页面

图 2-11　授权 DevEco Studio 访问华为账号

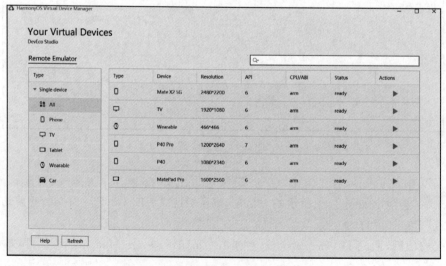

图 2-12　Virtual Device Manager 对话框

其中 Phone 有 Mate X2 5G、P40、P40 Pro，TV 就是华为智慧屏或者带有 HarmonyOS 的智能电视，Wearable 是华为智能手表，MatePad Pro 是华为平板。双击 P40 的三角形按钮，即可申请 P40 模拟器。

成功申请 P40 模拟器后，会显示一个视图，里面是 P40 预览器，如图 2-13 所示。该预览器与真机显示无异，真机上有的功能都有。在预览器的底部按钮用于控制预览器，如圆圈代表恢复初始界面，三角形代表返回。

需要注意的是，每次申请使用，只有 1 个小时（为了防止用户长时间占用模拟器），如果过期按之前的步骤重新申请即可。

2.2.3 运行 HelloWorld 工程

现在已经准备就绪，单击 IDE 上方工具栏中的三角形按钮（见图 2-14）运行 HarmonyOS App。

单击 IDE 工具栏的运行按钮后，运行效果如图 2-15 所示。

图 2-13　P40 模拟器

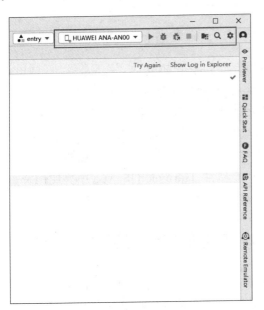

图 2-14　IDE 工具栏按钮

图 2-15　运行效果

2.2.4 分析 HelloWord 工程

基于 Java 开发的 HarmonyOS 工程的目录结构如图 2-16 所示。

下面介绍两个比较常用的目录：java 目录和 resources 目录。这两个目录的位置如下。

（1）java 目录：<HarmonyOS 工程根目录>/entry/src/main/java。

（2）resources 目录：<HarmonyOS 工程根目录>/entry/src/main/resources。

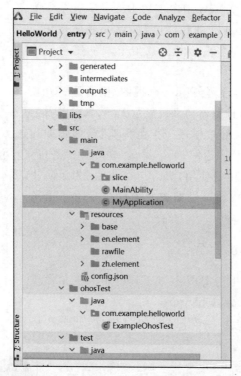

图 2-16　HarmonyOS 工程的目录结构

java 目录用于保存 HarmonyOS App 中的 Java 代码文件。resources 目录用来保存 HarmonyOS App 中的各种资源，如图像、字符串、数据库等。

打开 java 目录，会看到该目录中会有若干个 Java 包，如果是新创建的 HarmonyOS 工程，里面有 Ability 和 Application 两个 Java 文件，Ability 是应用所具备能力的抽象，也是应用程序的重要组成部分，一个应用可以具备多种能力（即可以包含多个 Ability）。Application 相当于是一个总的入口，然后去启动 Ability。如果 HarmonyOS App 中要使用资源，而且这些资源要被嵌入到 hap 文件中，则需要将这些文件放在 resources 目录下。

2.3　HarmonyOS 应用程序运行调试

调试是开发 App 的必备技能，任何一个比较复杂的 App 一般都需要多次调试才能成功，如果程序的运行效果与最初设计的效果不同，或者因为代码错误的原因导致程序异常中断停止，都需要查找原因，根据原因重新修改代码，直至效果与预期相同或者运行成功。

运行调试有多种方法，比较常用的两种调试方式是设置断点和输出日志。

2.3.1　设置断点

如果认为需要测试的代码或 Bug 在某行代码附近，可以单击该行代码前面的部分为该行设置断点，在这行代码前面会出现一个红点，如图 2-17 所示。

然后单击上方工具栏中如图 2-18 所示的"调试"按钮。

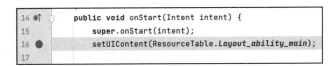

图 2-17　设置断点　　　　　　　　图 2-18　"调试"按钮

如果未遇到断点，单击"调试"按钮就会正常运行，如果遇到断点，程序就会在断点处停止运行，之后按 F7 或 F8 键，程序又可以继续进行，这样可以清楚地看到程序中相关变量的变化。F7 键代表 Step Into，Step Into 表示会跟踪到方法内部。当运行一个方法时，按 F7 键会继续跟踪方法内部，然后代码继续逐行运行。F8 键代表 Step Over，Step Over 会将方法当作一段代码整体运行，但不会跟踪到方法内部。所以，如果认为某一个方法没有问题，就按 F8 键，如果认为方法内部可能会有一些问题，就按 F7 键。

断点调试结果如图 2-19 所示，关于调试的功能按钮如表 2-2 所示。

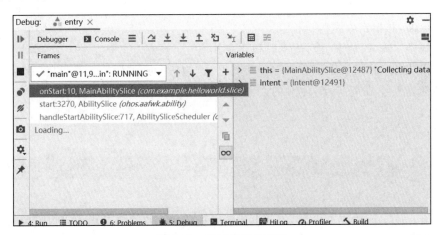

图 2-19　调试代码的过程

表 2-2　调试按钮

按钮	名称	快捷键	功能
▶	Resume Program	F9	当程序执行到断点时停止执行，单击此按钮程序继续执行
⤴	Step Over	F8	在单步调试时，直接前进到下一行（如果在函数中存在子函数时，不会进入子函数内单步执行，而是将整个子函数当作一步执行）
⤵	Step Into	F7	在单步调试时，遇到子函数后，进入子函数并继续单步执行
⤓	Force Step Into	Alt＋Shift＋F7	在单步调试时，强制下一步
⤒	Step Out	Shift＋F8	在单步调试执行到子函数内时，单击 Step Out 按钮会执行完子函数剩余部分，并跳出返回到上一层函数
↻	Rerun	Ctrl＋F5	重新启动调试
■	Stop	Ctrl＋F2	停止调试任务
⇥	Run To Cursor		断点执行到鼠标停留处，仅 TV、Wearable 支持

2.3.2 输出日志

HarmonyOS 提供了一种输出日志的方式，这就是 HiLog 类，该类提供了多个静态方法，用于输出不同级别的日志信息。这些静态方法如表 2-3 所示。

表 2-3 静态方法

方法名	功　能	方法名	功　能
debug	用于输出调试日志信息	fatal	用于输出致命错误日志信息
error	用于输出普通错误日志信息	warn	用于输出警告日志信息
info	用于输出普通的日志信息		

这 5 个方法的参数完全相同。

在输出日志之前，必须在 HiLogLabel 中定义日志类型、服务域和标记，应用于指定日志级别的界面，并指定隐私标识符。

日志级别：调试、信息、警告、错误和致命。

参数格式：printf 格式字符串，以％字符开头，包括参数类型标识符和变量参数。

隐私标识符：在每个参数中的％字符和参数类型标识符之间添加了{public}或{private}。

例如，error 方法的原型如下：

```
public static int error(HiLogLabel label, String format, Object... args);
```

参数 label：定义好的 HiLogLabel 标签。

参数 format：格式字符串，用于日志的格式化输出。格式字符串中可以设置多个参数，如格式字符串为"Failed to visit ％s."，"％s"是参数类型为 string 的变参标识，具体取值在 args 中定义。

使用输出日志方法的案例如下：

```
HiLogLabel label = new HiLogLabel(HiLog.LOG_App ,223, "MY_TAG");
HiLog.error(label, "这是一行错误信息,原因：%{private}s","URL 不可访问");
HiLog.warn(label,"这是一个警告,原因：%{public}s", "变量的值可能是负数");
```

HiLog.LOG_App：用于指定输出日志的类型。HiLog 中当前只提供了一种日志类型，即应用/服务日志类型 LOG_App。

223：用于指定输出日志所对应的业务领域，取值范围为 0x0～0xFFFFF，开发者可以根据需要进行自定义。

MY_TAG：用于指定日志标识，可以为任意字符串。

每个参数需添加隐私标识，分为{public}或{private}，默认为{private}。{public}表示日志打印结果可见；{private}表示日志打印结果不可见，输出结果为＜private＞。

参数 args：可以为 0 个或多个参数，是格式字符串中参数类型对应的参数列表。参数的数量、类型必须与格式字符串中的标识一一对应。执行这段代码，会在 HiLog 视图中输出如图 2-20 所示的日志信息。

```
02-27 13:24:06.541 5052-5052/com.example.myapplication E 000DF/MY_TAG: 这是一行错误信息，原因:<private>
02-27 13:24:06.541 5052-5052/com.example.myapplication W 000DF/MY_TAG: 这是一个警告，原因是：变量的值可能是负数
```

图 2-20　在 HiLog 视图中输出的日志信息

使用 error 方法会输出深红色的日志信息，日志的内容与其他方法输出的日志内容相同。不过要注意，使用 HiLog 的相关方法输出的日志分为 5 个级别，分别是 DEBUG（调试）、INFO（信息）、WARN（警告）、ERROR（错误）、FATAL（致命错误）。这 5 个级别分别用 5 个整数表示，这些整数都在 HiLog 类中定义，代码如下：

```java
public final class HiLog {
    public static final int DEBUG = 3;
    public static final int INFO = 4;
    public static final int WARN = 5;
    public static final int ERROR = 6;
    public static final int FATAL = 7;
}
```

如果要在 HiLog 视图中过滤这些级别的信息。只有不大于当前级别的信息才会显示。例如，要过滤 WARN 信息，只有 DEBUG、INFO 和 WARN 这 3 类信息才会被显示，由于 ERROR 和 FATAL 的值都比 WARN 大，所以这两类信息不会被显示。

2.4　HarmonyOS 应用程序设备调试

在实际开发过程中，需要在真机上测试后才能对外发布，有很多功能（如蓝牙、NFC、传感器等）在 HarmonyOS 模拟器上是无法测试的。下面会介绍在鸿蒙设备中如何调试鸿蒙应用程序。

2.4.1　生成签名文件

在真机上运行 App 与在模拟器上运行 App 不同。在真机上不管是调试还是发布，都需要 hap 文件进行签名。签名一个 hap 文件需要 4 类文件，即 p12、csr、cer 和 p7b 文件。其中 p12 文件可以自动生成，cer 和 p7b 文件需要到 AppGrallery Connect 去申请，然后下载。

1. 生成 key 和 CSR 文件

可选择"Build"→"Generate Key and CSR"菜单项生成相关文件，如图 2-21 所示。

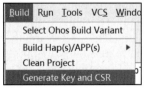

图 2-21　Build 工具栏

生成 Key 和 CSR 文件的具体步骤如图 2-22～图 2-27 所示。

图 2-22　Generate Key and CSR

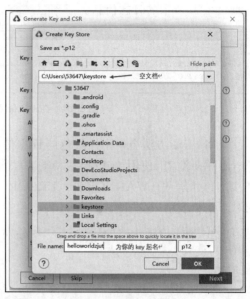

图 2-23　存放 Key 路径

图 2-24　设置密码图

图 2-25　填写详细信息

2. 配置调试证书

在华为的 AppGallery Connect 网站中配置项目的安全证书。

官网地址：https://developer.huawei.com/consumer/cn/service/josp/agc/index.html#/，单击"用户与访问"按钮，如图 2-28 所示。

3. 证书管理

调试应用需要申请证书，填写相关信息并使用之前生成的 CSR 文件进行申请即可，对

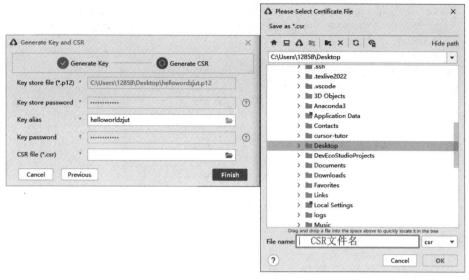

图 2-26　生成并保存 CSR 文件

图 2-27　CSR 生成成功

图 2-28　用户与访问

不同的应用需要申请不同的证书，如图 2-29 和图 2-30 所示。

填写证书信息，如图 2-30 所示。

图 2-29　证书管理

图 2-30　填写证书信息

4. 创建项目并添加应用

单击"我的项目"并创建项目，如图 2-31 所示。

项目创建完成后，还需在该项目下添加待调试的应用，进入项目设置页面，单击"添加应用"按钮，填写相关应用信息，需要注意的是，应用的包名是唯一的，网页提交信息中的应用包名与项目中 resources 目录下 config.json 文件中 bundlename 相同（如果包名已经存在就不能创建了），如图 2-32 所示。

5. 申请调试 Profile 证书

登录 AppGallery Connect 网站，在网站首页中单击"我的项目"，找到自己的项目，单击

图 2-31 创建项目

图 2-32 添加应用

创建的 HarmonyOS 应用,在左侧导航栏选择"HarmonyOS 应用"→"HAP Provision Profile 管理"命令,进入"管理 HAP Provision Profile"页面。

单击右上角"添加"按钮,在弹出的"HarmonyAppProvision 信息"窗口中添加 Profile,如图 2-33 所示。

6. 添加设备

打开 AppGallery Connect 页面,单击"用户与访问"按钮。

进入该界面后,在左侧导航栏选择"设备管理"选项,进入设备管理界面。然后单击右侧"添加设备"按钮,弹出"填写设备信息"对话框(见图 2-34),输入设备名称,选择类型,输入 UDID,然后单击"提交"按钮添加设备。

图 2-33　申请调试 Profile 证书

图 2-34　添加设备

如何获取 UDID：首先通过数据线连接手机，打开开发者模式，即在"设置"→"关于手机或关于平板电脑"中，连续多次单击"版本号"，直到提示"您正处于开发者模式"即可；然后在设置的"系统和更新"→"开发人员选项"中，打开"USB 调试"开关（注意所用的数据线必须是华为原装数据线）。

（1）使用 USB 方式，将 Phone 或者 Tablet 与 PC 端进行连接。

（2）在 Phone 或者 Tablet 中，USB 连接方式选择"传输文件"。

（3）在 Phone 或者 Tablet 中，会弹出"是否允许 USB 调试"的弹框，单击确定按钮。

在 DevEco Studio 进入 Termina 控制台，切到 HarmonyOS 的 SDK/toolchains 目录下，执行".\hdc.exe shell bm get -udid"命令（见图 2-35），即可获取到 UDID，如图 2-36 所示。

图 2-35　获取 UDID

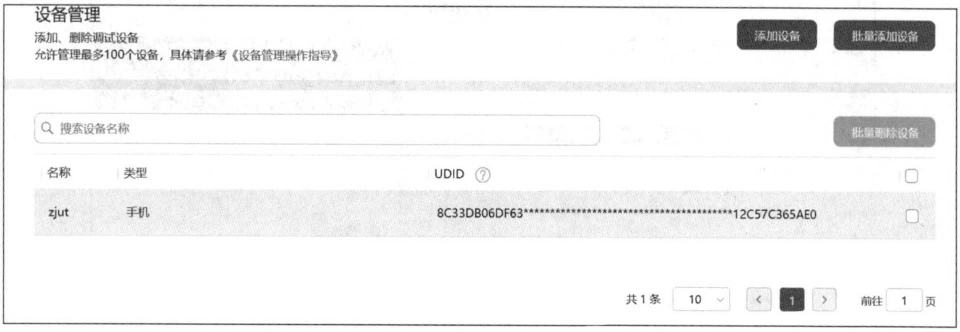

第 2 章　Hello World

图 2-36　成功添加设备

2.4.2　签名 HarmonyOS App

依次选择"File→Project Structure→Signing Configs"配置应用的签名信息,如图 2-37 所示。

图 2-37　签名 HarmonyOS App

设置完签名信息后,单击"OK"按钮进行保存,然后使用 DevEco Studio 生成 App。单击右上角的三角就可以真机调试了,如图 2-38 所示。

2.4.3　无线真机调试

使手机和电脑处于同一局域网下,先用数据线连接手机和计算机。通过以下命令打开 5555 端口。注意命令行路径在 Sdk/toolchains 下,hdc.exe 所在的目录,如图 2-39 所示。

图 2-38 效果图

图 2-39 命令端口

查看手机 IP 地址，选择"Tools"→"IP Connect"菜单项，如图 2-40 所示。

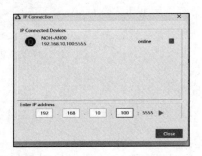

图 2-40 添加设备

至此完成了有线与无线连接方式的真机鸿蒙应用调试。

第 3 章 窗 口

Ability 是应用所具备能力的抽象,也是应用程序的重要组成部分,是 Harmony 应用程序的一大核心。一个应用可以包含多个 Ability,而 Ability 又可以分为 FA(Feature Ability)和 PA(Particle Ability)两种类型,FA 仅支持 Page Ability,它对用户是可见的,承载了一个业务可视化界面,即用户可通过 FA 与应用程序进行交互,PA 支持 Data Ability(第 9 章)和 Service Ability(第 10 章),它在后台运行,对用户是不可见的,PA 无法提供与用户交互的能力。本章主要对 Page Ability 进行讨论。

通过阅读本章,读者可以掌握:
- 如何使用 Page Ability。
- 如何在 Page Ability 之间进行交互。
- 如何进行跨设备迁移 Page Ability。
- 如何使用 AbilitySlice。
- Page Ability 与 AbilitySlice 的生命周期。

3.1 Page Ability 概述

Page Ability 是 FA 唯一支持的模板,它本质上是一个窗口,用于提供与用户交互的能力,类似于 Android 中的 Activity。另外,HarmonyOS 提供了 AbilitySlice,AbilitySlice 是指应用的单个页面控制逻辑的总和,相当于页面内的子窗口,类似于导航窗口,其功能与 Page Ability 相同,在切换时可以在同一个 Page Ability 内完成,也可以跳转至不同的 Page Ability。Page Ability 之间的切换相当于 Web 网页之间的切换,而 AbilitySlice 之间的切换相当于在一个 Web 页面下不同导航窗口之间的切换。

Page Ability 可以使用一个或多个 AbilitySlice,也可以不使用。在创建 HarmonyOS 工程时,包含了一个默认的 AbilitySlice(MainAbilitySlice.java)。当在一个 Page Ability 中使用多个 AbilitySlice 时,这些 AbilitySlice 所提供的功能之间应该具有高度的相关性,换言之,页面提供的功能之间有高度相关性时,应该在一个 Page Ability 下使用两个 AbilitySlice,而不必使用两个 Page Ability,以减少冗余。Page Ability 和 AbilitySlice 的关系如图 3-1 所示。

为了提高开发者的开发效率,使用 DevEco Studio 创建 HarmonyOS 工程时,IDE 提供了一些 Ability 模

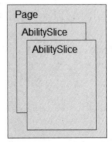

图 3-1　Page Ability 与 AbilitySlice 的关系

板供开发者使用,如图 3-2 所示,读者可以使用这些 Ability 模板快速生成一个 HarmonyOS 工程的框架,相当于一个简单的 Hello World 工程。

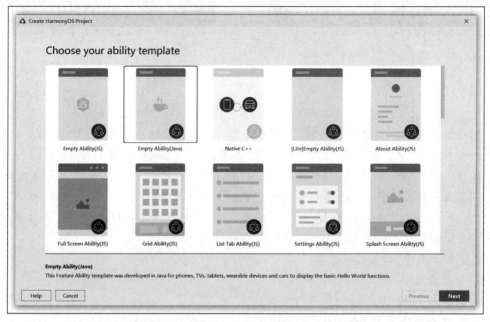

图 3-2　Ability 模板

3.2　Page Ability 的基本用法

为了提高开发效率,DevEco Studio 也为开发人员提供了自动创建 Page Ability 的功能,在创建过程中会自动创建 Page 类型的 Ability 类,同时创建一个 AbilitySlice 类以及布局文件,并自动向 config.json 文件中添加 Page Ability 的配置信息,这些都是开发工具自动完成的。为了使读者能够更好地理解和使用 Page Ability,本节将介绍如何手动创建 Page Ability、布局文件,以及如何装载布局文件,并在 config.json 中配置一个 Page Ability。

3.2.1　手动创建 Page Ability 类

Page Ability 是一个普通的 Java 类,其必须继承 Ability 类,该类属于 ohos.aafwk.ability 包,下面给出 MyFirstAbility 类的创建代码:

```
package com.example.createpageability;
import ohos.aafwk.ability.Ability;
public class MyFirstAbility extends Ability {

}
```

3.2.2　在 config.json 文件中注册 Page Ability

在 HarmonyOS App 中,所有 Ability 在使用前必须在 config.json 文件中的 abilities 配

置项配置。abilities 是一个对象数组,每一个对象表示一个 Ability,包括 Page Ability、Data Ability 和 Service Ability。MyFirstAbility 的配置代码如下:

```
{
    "skills": [
        {
            "actions": [
                "com.example.myfirstability"
            ]
        }
    ],
    "orientation": "unspecified",
    "formsEnabled": false,
    "name": "com.example.createpageability.MyFirstAbility",
    "icon": "$media:icon",
    "description": "$string:myfirstability_description",
    "label": "$string:myfirstability_label",
    "type": "page",
    "launchType": "standard"
}
```

下面对 abilities 配置信息中的常用属性进行介绍:

(1) skills 表示该 Ability 能够接收 Intent 的特征,是一个对象数组。

skills 中对象又有 3 种属性:

actions:表示能够接收的 Intent 的 action 值,可以包含一个或者多个 action;

entities:表示能够接收的 Intent 的 Ability 的类别,可以包含一个或者多个 entity;

uris:表示能够接收的 Intent 的 uri,可以包含一个或者多个 uri。

(2) orientation 表示该 Ability 的显示模式,该标签仅适用于 page 类型的 Ability。其只有 4 种取值:

unspecified:由系统自动判断显示方向;

landscape:横屏模式;

portrait:竖屏模式;

followRecent:跟随栈中最近的应用。

(3) formsEnabled 表示是否支持卡片功能,仅适用于 page 类型的 Ability。

(4) name 表示该 Ability 的名称,取值可采用反向域名方式表示,由包名加类名组成,如"com.example.createpageability.MyFirstAbility",也可以使用"."开头的类名方式表示,如".MyFirstAbility",一个应用中的每个 Ability 的 name 属性应该是唯一的。

(5) icon 表示 Ability 图标资源文件的索引,用于配置该 Ability 的图标,这里使用"$media:icon"来表示对 resources/base/media 下的 icon.png 文件的引用。

(6) description 表示对 Ability 的描述,这里使用"$string:myfirstability_description"来表示对 resources/base/element 下的 string.json 文件中定义的字符串的引用,以后所有的字符串都将在这个文件中定义,使用"$string:"来引用。

(7) label 表示 Ability 对用户显示的名称。

（8）type 表示该 Ability 的类型。

其一共有四种取值：

page：表示基于 Page 模板开发的 FA，用于提供与用户交互的能力；

service：表示基于 Service 模板开发的 PA，用于提供后台运行任务的能力；

data：表示基于 Data 模板开发的 PA，用于对外部提供统一的数据访问抽象；

CA：表示支持其他应用以窗口方式调起该 Ability。

（9）launchType 表示 Ability 的启动模式。

它支持"standard""singleMission"和"singleton"3 种模式：

standard：表示该 Ability 可以有多实例，"standard"模式适用于大多数应用场景；

singleMission：表示该 Ability 在每个任务栈中只能有一个实例；

singleton：表示该 Ability 在所有任务栈中仅可以有一个实例。例如，具有全局唯一性的呼叫来电界面即采用"singleton"模式。该标签仅适用于手机、平板、智慧屏、车机、智能穿戴。

（10）reqPermissions：表示此 Ability 需要申请的权限。

3.2.3 创建布局文件

HarmonyOS App 所有的布局文件将放在 entry/src/main/resources/base/layout 目录下面，首先，创建一个 Layout Resource File 并命名为 my_first_layout.xml，然后输入以下代码：

```xml
<?xml version="1.0" encoding="utf-8"?>
<DirectionalLayout
    xmlns:ohos="http://schemas.huawei.com/res/ohos"
    ohos:height="match_parent"
    ohos:width="match_parent"
    ohos:alignment="center"
    ohos:orientation="vertical">
    <Text
        ohos:id="$+ id:text"
        ohos:height="200fp"
        ohos:width="match_parent"
        ohos:text_alignment="center"
        ohos:text="$string:myfirstability_description"
        ohos:text_size="30fp"></Text>
    <Button
        ohos:id="$+ id:button"
        ohos:height="50fp"
        ohos:width="200fp"
        ohos:background_element="#00ff00"
        ohos:text="销毁 Ability"
        ohos:text_size="30vp"></Button>
</DirectionalLayout>
```

关于布局和组件，之后的章节会详细进行介绍，读者在此模仿例子使用即可，这里用到的是方向布局 DirectionalLayout，也是创建布局文件时，IDE 会创建的默认布局。这里用到了两个组件：Text 和 Button。

3.2.4 静态装载布局文件

创建完布局文件后，需要将布局文件加载到 Page Ability 上，才能显示布局中的组件。通常需要在 Page Ability 启动时装载布局文件，也就是需要在 Page Ability 的生命周期方法中的 onStart() 方法里完成。关于生命周期方法，后面小节会详细介绍，读者只需要知道 onStart() 方法在 Page Ability 启动时调用即可，通常会在这个方法里做一些初始化的工作，例如，加载布局文件，初始化组件，为组件添加事件监听器等。

现在需要调用父类 Ability 的 onStart() 方法，并使用 super.setUIContent() 方法加载之前创建的布局文件 my_first_layout.xml，注意在写 Java 代码时，用到了其他类或其他类的静态方法时，需要 import 特定的类，这一点在之后的代码中不做体现，代码如下：

```
import ohos.aafwk.ability.Ability;
import ohos.aafwk.content.Intent;
public class MyFirstAbility extends Ability {
    //调用父类 onStart()方法
    public void onStart(Intent intent)
    {
        super.onStart(intent);
        super.setUIContent(ResourceTable.Layout_my_first_layout);
    }
}
```

在 HarmonyOS App 中，系统会将所有静态资源与一个 int 类型的值进行绑定，并将这些值以常量的形式定义在 static 类型的类 ResourceTable 中，以便通过这个静态类调用这些值，通过该值引用相关资源。这些值是自动生成的，以资源文件的名称加上资源类型(作为前缀)作为变量名。例如，布局文件生成的 ID 需要加上前缀 Layout，本例的布局文件是 my_first_layout.xml，因此在 ResourceTable 类中会自动生成 ID：Layout_my_first_layout。根据这个生成规则，还要求资源文件的命名必须符合 Java 标识符的命名规则，否则无法在 ResourceTable 自动生成 ID。

3.2.5 显示 Page Ability

目前为止，一个小型但完整的 Page Ability 已经创建完成，最后一步就是显示这个创建好的 Page Ability。如果想让 MyFirstAbility 作为应用的主 Ability(即程序运行后的显示的第一个页面)显示，可以修改 MyFirstAbility 的配置信息的 skills 部分，将其修改为如下形式：

```
"skills": [
    {
        "entities": [
            "entity.system.home"
```

```
        ],
        "actions": [
            "action.system.home"
        ]
    }
]
```

需要注意的是,一个应用只能有一个主 Ability,但在 config.json 文件中还有其他 Ability 的 actions 也设为"action.system.home",而 HarmonyOS 只会显示在 config.json 文件中遇到的第一个主 Ability。因此如果需要将您的 Ability 设为主 Ability,就需要将您的 Ability 配置信息作为 abilities 中的第一个元素,或者删除其他的 action 属性值为"action.system.home"的配置项。将 MyFirstAbility 作为主 Ability 显示时,打开应用会看到如图 3-3(左)所示的页面,而如果从其他页面显示 MyFirstAbility,如从图 3-3(右)显示 MyFirstAbility,显示效果和图 3-3(左)相同。

图 3-3　显示 **MyFirstAbility**

3.2.6　销毁 Page Ability

在 Page Ability 使用完后,需要关闭(或销毁)Page Ability,调用如下代码即可销毁 Page Ability:

```
terminateAbility();
```

该方法属于 Ability 类,如果在 AbilitySlice(后面章节介绍)中需要调用该方法,需要获得包含该 AbilitySlice 的 Ability 对象。

3.3 Page Ability 之间的交互

本小节将介绍两个不同的 Page Ability 之间如何进行交互，例如，在两个不同 Page Ability 之间传递数据、通过显式和隐式的方式在一个 Page Ability 中显示另一个 Page Ability。

3.3.1 Intent 的基本概念

Intent 是对象之间传递信息的载体。例如，当一个 Ability 需要启动另一个 Ability 时，或者一个 AbilitySlice 需要导航到另一个 AbilitySlice 时，可以通过 Intent 指定启动的目标同时携带相关数据。Intent 的构成元素包括 Operation 与 Parameters。

Operation 又包含以下 7 项属性：

（1）Action：表示动作，可以使用系统内置的 Action，也可以使用用户在 config.json 文件中自定义的 Action。

（2）Entity：表示类别，可以使用系统内置的 Action，也可以使用用户在 config.json 文件中自定义的 Action。

（3）URI：表示 URI 描述。如果在 Intent 中指定了 URI，则 Intent 将匹配指定的 URI 信息。

（4）Flags：表示处理 Intent 的方式。如 Intent.FLAG_ABILITY_CONTINUATION 标记在本地的一个 Ability 是否可以迁移到远端设备继续运行。

（5）BundleName：表示包描述。

（6）AbilityName：表示待启动的 Ability 名称。如果在 Intent 中同时指定了 BundleName 和 AbilityName，则 Intent 可以直接匹配到指定的 Ability。

（7）DeviceId：表示运行指定 Ability 的设备 ID。

Parameters 是一种支持自定义的数据结构，开发者可以通过 Parameters 传递某些请求所需的额外信息。

3.3.2 显式使用 Intent

所谓显式使用就是同时指定 Operation 属性的 BundleName 和 AbilityName 两项子属性，即根据 Ability 的全称启动 Ability，隐式使用就是指未同时指定 Operation 属性的 BundleName 和 AbilityName，根据 Operation 的其他属性启动 Ability，通常指定 Ability 指定的 action 属性启动 Page Ability。

【例 3.1】 首先需要创建一个名为 ability_main.xml 的布局文件，同时添加两个 Button 组件并绑定到 MainAbility，主 Ability 用于演示显式使用 Intent 和隐式使用 Intent 的区别。布局文件代码如下：

```
<?xml version="1.0" encoding="utf-8"?>
<DirectionalLayout
    xmlns:ohos="http://schemas.huawei.com/res/ohos"
    ohos:height="match_parent"
```

```
            ohos:width="match_parent"
            ohos:alignment="center"
            ohos:orientation="vertical">
    <Button
        ohos:id="$+ id:explicit"
        ohos:height="40fp"
        ohos:width="200fp"
        ohos:text="Explicit"
        ohos:text_size="25fp"
        ohos:background_element="$graphic:background_btn"></Button>
    <Button
        ohos:id="$+id:implicit"
        ohos:height="40fp"
        ohos:width="200fp"
        ohos:top_margin="20fp"
        ohos:text="Implicit"
        ohos:text_size="25fp"
        ohos:background_element="$graphic:background_btn"></Button>
</DirectionalLayout>
```

然后创建一个名为 ExplicitAbility 的 Page Ability,同时绑定一个布局文件,并在配置文件中进行注册,由于代码简单,因此不在此赘述。显式地显示 Page Ability 的步骤如下:

(1) 创建 Intent 对象;

(2) 使用 Intent.OperationBuilder 类构造包含 BundleName 与 AbilityName 的 Operation 对象;

(3) 通过 Intent 的 setOperation()方法指定 Intent 的 Operation 对象;

(4) 调用 startAbility()方法显示 Page Ability。

按照以上步骤编写显示 ExplicitAbility 的代码如下:

```
Intent intent = new Intent();
Operation operation = new Intent.OperationBuilder()
    //指定设备标识,空串表示当前设备
    .withDeviceId("")
    //指定包名
    .withBundleName("com.example.explicitintent")
    //指定 Page Activity 的 name 属性值
    .withAbilityName("com.example.explicitintent.ExplicitIntentAbility")
    .build();
intent.setOperation(operation);
startAbility(intent);
```

其中,withDeviceID 用于指定设备 ID,在 config.json 文件中的 deviceConfig 中进行配置,空串表示当前设备。withBundleName 指定的是 HarmonyOS App 的包名,在 config.json 文件中 bundlename 属性指定。withAbilityName 指定的是 Page Ability 的全名(包名+类名)。最后需要调用 build()方法返回一个 Operation 对象。执行程序,单击"Explicit"按钮将会显示 ExplicitAbility,效果如图 3-4 所示。

第 3 章　窗口

图 3-4　显式使用 Intent

3.3.3　隐式使用 Intent

首先创建一个名为 Implicit1Ability 的 Page Ability 类，同时绑定一个布局文件，然后在 config.json 文件中添加如下配置信息：

```json
{
    "skills": [
        {
            "actions": [
                "action.implicit"
            ]
        }
    ],
    "orientation": "unspecified",
    "name": "com.example.intentuse.Implicit1Ability",
    "icon": "$media:icon",
    "description": "$string:implicit1",
    "label": "$string:implicit1",
    "type": "page",
    "launchType": "standard"
}
```

由于在隐式使用 Intent 的方式中需要用到 action 属性，因此需要配置 action 为"action. implicit"，并通过如下代码显示 Implicit1Ability：

```
Intent intent = new Intent();
Operation operation = new Intent.OperationBuilder()
                                 .withAction("action.implicit ")
                                 .build();
intent.setOperation(operation);
startAbility(intent);
```

隐式使用 Intent 时,不需要指定 BundleName 与 AbilityName,只需要指定 Action 即可。注意,由于在实际开发过程中,有可能遇到多个 Page Ability 指定同一个 action,因此,使用隐式显示 Page Ability 时,HarmonyOS 会弹出一个列表,列表中是所有绑定了同一个 action 的 Page Ability,供用户选择到底使用哪一个 Page Ability。这也是和显式地显示 Page Ability 的一个重要区别,由于显式方法指定确定的 BundleName 与 AbilityName,因此会显示指定 Page Ability。下面通过一个例子介绍多个 Page Ability 指定同一个 action 时会出现的情况。

再创建一个名为 Implicit2Ability 的 Page Ability,同时绑定布局文件,在 config.json 文件中的配置信息如下所示:

```
{
    "skills": [
        {
            "actions": [
                "action.implicit"
            ]
        }
    ],
    "orientation": "unspecified",
    "name": "com.example.intentuse.Implicit2Ability",
    "icon": "$media:icon",
    "description": "$string:implicit2",
    "label": "$string:implicit2",
    "type": "page",
    "launchType": "standard"
}
```

显然,这两个 Page Ability 指定了同一个 action,使用如下代码显示 Page Ability:

```
Intent intent = new Intent();
Operation operation = new Intent.OperationBuilder()
    .withAction("action.implicit ")
    .build();
    intent.setOperation(operation);
startAbility(intent);
```

执行程序,单击"Implicit"按钮执行以上代码,系统将会弹出一个列表供用户选择进入哪个 Page Ability,如图 3-5 所示。

图 3-5　隐式使用 Intent

3.3.4　Page Ability 之间的通信

一般来说，Page Ability 之间是需要进行通信的，而通信方式可以分为两种，一种是 Page Ability 向下一个 Page Ability 传递数据；另一种是 Page Ability 向上一个 Page Ability 返回数据。

第一种通信方式是单向的，因此只用通过 Intent 对象的 setParam()函数携带相关数据并调用 startAbility()方法传递到下一个 Page Ability 即可，代码如下：

```
Intent intent = new Intent();
intent.setParam("data","mydata");
startAbility(intent);
```

执行这段代码，就能将数据传到下一个 Page Ability，在下一个 Page Ability 中，通过调用 Intent.getStringParam()方法就能获得传递过来的数据，代码如下：

```
Intent myintent = getIntent();
String data = myintent.getStringParam("data");
```

第二种通信方式是双向的，从 Page Ability1 传递数据并跳转到 Page Ability2 时，需要 Page Ability2 返回数据给 Page Ability1。此时，需要使用 startAbilityForResult()方法。当使用该方法跳转到 Page Ability2 时，由 Page Ability2 返回到 Page Ability1 后，Page Ability1 会自动调用 onAbilityResult()方法，因此，需要提前在 Page Ability1 中重写该方法用于接收从 Page Ability2 传递过来的数据，该方法的原型：

```
protected void onAbilityResult (int requestCode, int resultCode,
                    Intent resultData);
```

onAbilityResult()方法的参数含义如下。

(1) requestCode：请求码，通过在启动 Ability 时设置，即 startAbilityResult()方法的第 2 个参数。

(2) resultCode：响应码，在 Page Ability2 中调用 setResult()方法时设置。

(3) resultData：由 Page Ability2 返回的 Intent 对象形式的数据。

这里重点介绍一下 requestCode 和 resultCode。因为在 Page Ability 中可能会通过 startAbilityForResult()方法显示多个 Page Ability，所以 onAbilityResult()方法可能是多个 Page Ability 共享的，这就要求在 onAbilityResult()方法中区别是哪一个 Page Ability 返回的结果。可以通过 requestCode 或 resultCode 单独区分不同的 Page Ability，也可以使用 requestCode 和 resultCode 共同区分不同的 Page Ability。如果在 Page Ability1 中通过 startAbilityForResult()方法显示 Page Ability2，那么 requestCode 应该在 Page Ability1 中指定，而 resultCode 应该在 Page Ability2 中指定。

【例 3.2】演示两个 Page Ability 之间是如何通信的。在 Page Ability1 调用 startAbilityForResult()方法显示 Page Ability2，并传递一个字符串类型数据，Page Ability2 接收这个数据，并显示在页面上，然后关闭 Page Ability2 并返回一些数据给 Page Ability1，Page Ability1 在 onAbilityResult()方法中接收来自 Page Ability2 的数据。

首先创建一个名为 pageability1_layout.xml 的布局文件，布局中包括一个 Text，用于显示 Page Ability2 返回的数据，一个 Button，用于显示 Page Ability2，代码如下：

```xml
<?xml version="1.0" encoding="utf-8"?>
<DirectionalLayout
    xmlns:ohos="http://schemas.huawei.com/res/ohos"
    ohos:height="match_parent"
    ohos:width="match_parent"
    ohos:alignment="center"
    ohos:orientation="vertical">
    <Text
        ohos:id="$+id:text_page1"
        ohos:height="50fp"
        ohos:width="match_parent"
        ohos:text_alignment="center"
        ohos:text="待接收数据..."
        ohos:text_size="25fp"></Text>
    <Button
        ohos:id="$+id:button_page1"
        ohos:height="50fp"
        ohos:width="300fp"
        ohos:background_element="$graphic:bcakground_btn"
        ohos:text="To PageAbility2"
        ohos:text_size="30fp"></Button>
</DirectionalLayout>
```

然后创建一个名为 pageability2_layout.xml 的布局文件，布局中包括一个 Text，用于显示 Page Ability1 传递来的数据，一个 Button，用于关闭 Page Ability2，代码如下：

```xml
<?xml version="1.0" encoding="utf-8"?>
<DirectionalLayout
    xmlns:ohos="http://schemas.huawei.com/res/ohos"
    ohos:height="match_parent"
    ohos:width="match_parent"
    ohos:alignment="center"
    ohos:orientation="vertical">
    <Text
        ohos:id="$+id:text_page2"
        ohos:height="50fp"
        ohos:width="match_parent"
        ohos:text_alignment="center"
        ohos:text="待接收数据..."
        ohos:text_size="25fp"></Text>
    <Button
        ohos:id="$+id:button_page2"
        ohos:height="50fp"
        ohos:width="300fp"
        ohos:background_element="$graphic:bcakground_btn"
        ohos:text="Back PageAbility1"
        ohos:text_size="30fp"></Button>
</DirectionalLayout>
```

接下来创建一个名为 PageAbility1 的 Page Ability 类，主要代码如下：

```java
public class PageAbility1 extends Ability {
    Button button;
    Text text;
    public void onStart(Intent intent){
        super.onStart(intent);
        //加载布局文件
        super.setUIContent(ResourceTable.Layout_pageability1_layout);
        //通过组件 ID 获取组件
        button = (Button)findComponentById(ResourceTable.Id_button_page1);
        text = (Text)findComponentById(ResourceTable.Id_text_page1);
        //绑定按钮监听事件
        button.setClickedListener(new Component.ClickedListener() {
            @Override
            public void onClick(Component component) {
                Intent intent = new Intent();
                //传递一个字符串
                intent.setParam("pageability1_data",
                    "这是 Page Ability1 传递的数据");
                Operation operation = new Intent.OperationBuilder()
                    .withBundleName("com.example.interactionofpage")
                    .withAbilityName("com.example.interactionofpage
                    .PageAbility2").build();
                intent.setOperation(operation);
                //显示 PageAbility2,设置请求码为 100
                startAbilityForResult(intent,100);
```

```java
            }
        });
    }
    //重写onAbilityResult方法,当PageAbility2关闭时自动调用
    protected void onAbilityResult(int requestCode, int resultCode,
                            Intent resultData){
        switch (requestCode){
            case 100:
                switch(resultCode){
                    case 101:
                        //接收返回的数据
                        String data = resultData.getStringParam
                        ("pageability2_data");
                        //将接收到的数据显示在页面中Text组件上
                        if(data!=null)text.setText(data);
                        break;
                }
                break;
        }
    }
}
```

这段代码实现了使用 startAbilityForResult() 方法从 PageAbility1 显示 PageAbility2，并传递一个字符串类型数据，重写了 onAbilityResult()，对 requestCode 和 resultCode 进行了判断，如果满足条件，则接收返回的数据，现在需要创建一个名为 PageAbility2 的 PageAbility 类，主要代码如下：

```java
public class PageAbility2 extends Ability {
    Button button;
    Text text;
    public void onStart(Intent intent){
        super.onStart(intent);
        //加载布局文件
        super.setUIContent(ResourceTable.Layout_pageability2_layout);
        //获取PageAbility1传递的数据
        Intent myintent = getIntent();
        String data = myintent.getStringParam("pageability1_data");
        //通过组件ID获取组件
        button = (Button)findComponentById(ResourceTable.Id_button_page2);
        text = (Text)findComponentById(ResourceTable.Id_text_page2);
        //将传递来的数据显示在页面的Text组件上
        if(data != null)text.setText(data);
        //绑定按钮监听,返回PageAbility1,并返回一个字符串类型数据
        button.setClickedListener(new Component.ClickedListener() {
            @Override
            public void onClick(Component component) {
                Intent resultintent = new Intent();
                //设置返回到PageAbility1的数据
                resultintent.setParam("pageability2_data",
                        "这是PageAbility2返回的数据");
```

```
                //设置响应码以及返回的Intent对象
                setResult(101, resultintent);
                //关闭PageAbility2
                terminateAbility();
            }
        });
    }
}
```

现在,两个Page Ability已经准备就绪,最后在config.json文件中注册这两个Page Ability即可使用这两个Page Ability。配置信息如下:

```
{
    "orientation": "unspecified",
    "name": "com.example.interactionofpage.PageAbility1",
    "icon": "$media:icon",
    "description": "$string:pageability1",
    "label": "$string:pageability1",
    "type": "page",
    "launchType": "standard"
},
{
    "orientation": "unspecified",
    "name": "com.example.interactionofpage.PageAbility2",
    "icon": "$media:icon",
    "description": "$string:pageability2",
    "label": "$string:pageability2",
    "type": "page",
    "launchType": "standard"
}
```

运行程序,效果如图3-6所示。

图3-6　Page Ability交互效果图

3.4 Page Ability 的启动类型

在之前的例子中,所有 Ability 的配置信息 launchType 的值都为 standard,这也是 launchType 属性的默认值。launchType 还有两个值为 singleMission 和 singleton,本节将对这 3 种属性值的作用进行介绍。

standard 表示此 Ability 可以创建多个实例,且在任何情况下,无论 Page Ability 被显示多少次,每次被显示都会创建一个新的 Page Ability 实例。

singleMission 表示此 Ability 在每个任务栈中只能有一个实例。如果要显示的 Page Ability 在栈顶,那么再次显示这个 Page Ability 时,不会再创建新的 Page Ability 实例,而是直接使用这个 Page Ability 实例。如果 Page Ability 上面有其他的 Page Ability,那么首先弹出这些 Page Ability,然后再复用这个 Page Ability。

singleton 表示该 Ability 在所有任务栈中只能有一个实例,例如,具有全局唯一性的呼叫来电界面即采用"singleton"模式。

其中涉及的栈是 HarmonyOS 管理 Page Ability 的模式。由于 HarmonyOS App 只能显示一个 Page Ability,HarmonyOS App 会使用栈来管理 App 中的所有 Page Ability,只有在栈顶的 Page Ability 才能显示,如果要让非栈顶的 Page Ability 显示,那么就需要将要显示的 Page Ability 之前的所有 Page Ability 销毁,使得要显示的 Page Ability 位于栈顶,要销毁 Page Ability,则需要调用 terminateAbility()方法。

【例 3.3】 由于 singleMission 模式和 singleton 模式都只允许 Ability 只有一个实例,本例仅演示 standard 和 singleMission 启动类型的区别。

首先创建一个名为 LaunchTypeAbility1 的 Page Ability 类,主要代码如下:

```java
public class LaunchTypeAbility1 extends Ability {
    private static int count = 0;        //计数器
    public void onStart(Intent intent){
        super.onStart(intent);
        super.setUIContent(ResourceTable.Layout_launchtype);
        //每创建一个 LaunchTypeAbility1 对象,计数器加一
        count++;
        Button button = (Button)findComponentById(ResourceTable.Id_button1);
        Text count1 = (Text)findComponentById(ResourceTable.Id_count1);
        if(count1 != null){
            //将计数器的值显示在页面中
            count1.setText(String.valueOf(count));
        }
        //设置按钮单击事件监听器
        button.setClickedListener(new Component.ClickedListener() {
            @Override
            public void onClick(Component component) {
                Intent intent = new Intent();
                Operation operation = new Intent.OperationBuilder()
                    .withBundleName("com.example.launchtype").withAbilityName
```

```
                    ("com.example.launchtype.LaunchTypeAbility1").build();
                intent.setOperation(operation);
                startAbility(intent);
            }
        });
    }
}
```

其次在 config.json 文件中添加这个类的配置信息，注意，先将这个类的 launchType 属性都设置为 standard，在 LaunchTypeAbility1 重复显示 LaunchTypeAbility1，这样可以很好地观察到在 standard 和 singleMission 这两种启动模式下，计数器的变化情况。

LaunchTypeAbility1 这个类中定义了一个变量 count 作为计数器，每次创建一个 LaunchTypeAbility1 实例对象，计数器都会加一，在 standard 启动模式下，显示三次 LaunchTypeAbility1，系统则会创建三个 LaunchTypeAbility1 实例对象，因此计数器的值变为 3，页面效果如图 3-7 所示。

现在将 LaunchTypeAbility1 的启动模式改为 singleMission，然后再显示三次 LaunchTypeAbility1。由于在 singleton 模式下，显示 LaunchTypeAbility1 并不会创建新的 LaunchTypeAbility1 实例对象，可以观察到计数器的值都为 1，页面效果如图 3-8 所示。

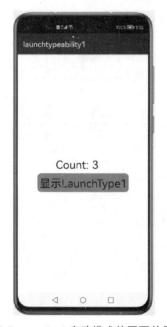
图 3-7　standard 启动模式的页面效果图

图 3-8　singleMission 启动模式的页面效果

3.5　Page Ability 的跨设备迁移

HarmonyOS 的一大技术特性是分布式任务调度，所谓分布式任务调度，就是构建统一的分布式服务管理（发现、同步、注册、调用）机制，支持对跨设备的应用进行远程启动、远程调用、远程连接以及迁移等操作,能够根据不同设备的能力、位置、业务运行状态、资源使用情况,以及用户的习惯和意图,选择合适的设备运行分布式任务。而 Page Ability 的跨设备迁移则依赖了分布式设备调度中的业务迁移能力,是分布式任务调度的一个具体实现。跨设备迁移支持将 Page 在同一用户的不同设备间迁移,以便支持用户无缝切换的诉求。实现这个操作的前提是参加跨设备迁移的设备在同一个网段内,或者登录了同一个 HUAWEI 账号,接下来将介绍跨设备前需要准备的工作。

3.5.1 跨设备迁移前的准备工作

在进行跨设备迁移之前(后面几章所讲的跨设备调用 Data Ability、Service Ability 也同样要进行这些准备),需要对 HarmonyOS 设备做一些准备工作:

(1)打开 HarmonyOS 设备的蓝牙,并把设备名称修改为可识别的名称,如图 3-9 所示;

(2)将 HarmonyOS 设备连入 Wi-Fi,并且所有参与跨设备迁移的 HarmonyOS 设备在同一个网段;

(3)所有参与跨迁移的 HarmonyOS 设备登录同一个 HUAWEI 账号;

(4)选择"设置"→"超级终端"查看附近设备如图 3-10 所示。

图 3-9　设置设备名称

图 3-10　附近设备

3.5.2 获取设备列表

跨设备迁移需要知道目标设备的 ID,因此需要提前获取所有可用的设备 ID。HarmonyOS 提供了一个用于获取所有设备信息的方法,即 DeviceManager.getDeviceList()方法,该方法返回一个 List 列表,类型是 DeviceInfo。DeviceInfo 类型描述了设备的名称和设备 ID 等相关信息，实现代码如下：

```
List<DeviceInfo> deviceInfoList =
        DeviceManager.getDeviceList(DeviceInfo.FLAG_GET_ONLINE_DEVICE)
```

其中，getDeviceList()方法的参数表示获取哪种状态下的设备的信息，一共有 3 种取值：

(1) DeviceInfo.FLAG_GET_ONLINE_DEVICE：所有在线设备；

(2) DeviceInfo.FLAG_GET_OFFLINE_DEVICE：所有离线设备；

(3) DeviceInfo.FLAG_GET_ALL_DEVICE：所有设备。

一般会使用第一个值，因为只有设备在线，才能进行迁移 Page Ability 操作。

【例 3.4】 实现一个获取在线设备列表的 Page Ability，单击一个设备会返回该设备的 ID。首先创建一个名为 device_ids.xml 的布局文件，代码如下：

```
<?xml version="1.0" encoding="utf-8"?>
<DirectionalLayout
    xmlns:ohos="http://schemas.huawei.com/res/ohos"
    ohos:height="match_parent"
    ohos:width="match_parent"
    ohos:orientation="vertical">
    <Text
        ohos:height="50vp"
        ohos:width="match_parent"
        ohos:text="可用设备 ID"
        ohos:text_size="35vp"></Text>
    <ListContainer
        ohos:id="$+id:device_id_List"
        ohos:height="match_parent"
        ohos:width="match_parent"></ListContainer>
</DirectionalLayout>
```

布局中包含了一个 ListContainer 组件，用于显示所有设备信息，再创建一个名为 device_id_item.xml 的布局文件，作为 ListContainer 组件的 Item 布局，代码如下：

```
<?xml version="1.0" encoding="utf-8"?>
<DirectionalLayout
    xmlns:ohos="http://schemas.huawei.com/res/ohos"
    ohos:height="match_parent"
    ohos:width="match_parent"
    ohos:orientation="vertical">
    <DirectionalLayout
        ohos:height="50vp"
        ohos:width="match_parent"
        ohos:orientation="horizontal">
        <Text
            ohos:height="match_parent"
            ohos:width="150vp"
            ohos:text="device_name:"
```

```xml
        ohos:text_size="20fp"></Text>
    <TextField
        ohos:id="$+id:device_name"
        ohos:height="match_parent"
        ohos:width="match_parent"
        ohos:text="huawei"
        ohos:text_size="20vp"></TextField>
</DirectionalLayout>
<DirectionalLayout
    ohos:height="200vp"
    ohos:width="match_parent"
    ohos:orientation="horizontal">
    <Text
        ohos:height="match_parent"
        ohos:width="150vp"
        ohos:text="device_id:"
        ohos:text_size="20fp"></Text>
    <TextField
        ohos:id="$+id:device_id"
        ohos:height="match_parent"
        ohos:width="match_parent"
        ohos:text="000"
        ohos:text_size="20vp"></TextField>
</DirectionalLayout>
</DirectionalLayout>
```

该布局文件中放置的两个 TextField 组件,分别显示设备名称和设备 ID。接下来是 Page Ability 的实现。创建一个名为 DeviceIDAbility 的 Page Ability 类,代码如下:

```java
public class DeviceIDAbility extends Ability {
    //存放获取到的在线设备信息
    private List<DeviceInfo> deviceInfoList;
    //展示设备信息的容器
    private ListContainer listContainer;
    //获取所有在线设备信息
    private static List<DeviceInfo> getAllOnlineDeviceInfo(){
        List<DeviceInfo> deviceInfoList =
            DeviceManager.getDeviceList(DeviceInfo.FLAG_GET_ONLINE_DEVICE);
        if(deviceInfoList == null || deviceInfoList.size() == 0)
        {
            return new ArrayList<>();
        }
        else{
            return deviceInfoList;
        }
    }
    public void onStart(Intent intent) {
```

```java
        super.onStart(intent);
        super.setUIContent(ResourceTable.Layout_device_ids);
        deviceInfoList = getAllOnlineDeviceInfo();
        listContainer= (ListContainer) findComponentById
            (ResourceTable.Id_device_id_List);
    //设置一个Provider
    listContainer.setItemProvider(new BaseItemProvider() {
        @Override
        public int getCount() {
            return deviceInfoList.size();
        }
        @Override
        public Object getItem(int i) {
            return deviceInfoList.get(i);
        }
        @Override
        public long getItemId(int i) {
            return i;
        }
        @Override
    public Component getComponent(int i, Component component,
            ComponentContainer componentContainer) {
        if(component == null)
        {//若没有指定列表项布局,则从资源列表加载一个布局文件
            component = (DirectionalLayout)LayoutScatter
                .getInstance(DeviceIDAbility.this)
                .parse(ResourceTable.Layout_device_id_item, null, false);
        }
        Text devicename = (Text)component.findComponentById
            (ResourceTable.Id_device_name);
        Text deviceid =(Text) component.findComponentById
            (ResourceTable.Id_device_id);
        if(devicename != null)
        {
            //显示设备名
            devicename.setText(deviceInfoList.get(i).getDeviceName());
        }
        if(deviceid != null)
        {
            //显示设备ID
            deviceid.setText(deviceInfoList.get(i).getDeviceId());
        }
        return component;
    }
});
//为ListContainer的所有Item项设置事件监听器
f(listContainer != null){
```

```
            listContainer.setItemClickedListener(new ListContainer
            .ItemClickedListener() {
                @Override
                public void onItemClicked(ListContainer listContainer,
                Component component, int i, long l) {
                    //当单击某个 Item 后,将该 Item 所对应的设备 ID 返回
                    String device_id = deviceInfoList.get(i).getDeviceId();
                    Intent resultIntent = new Intent();
                    //返回设备 ID
                    resultIntent.setParam("deviceID", device_id);
                    //设置响应码并携带 Intent 类型的数据
                    setResult(101, resultIntent);
                    terminateAbility();
                }
            });
        }
    }
}
```

在 DeviceIDAbility 类中为 ListContainer 组件装载列表项时,需要用到 LayoutScatter 类的 parse()方法,该方法的原型如下:

```
public Component parse(int xmlId, ComponentContainer root,
                       boolean attachToRoot)
```

parse()方法的参数解释如下:

（1）xmlId：需要装载的布局文件 Id；

（2）root：指定需要装载布局文件的根组件；

（3）attachToRoot：若不想指定根组件,该值可以设为 false。

parse()方法返回的是 Component 类型的对象,利用 getComponent()方法的第二个参数 component,可以通过该参数传入列表项的根视图,即利用布局文件 device_id_item.xml 创建的列表项视图,之后若需获取到列表项视图中的组件并操作布局中的组件时,只需利用 component.findComponentById()方法即可。

然后在 MainAbilitySlice 类中利用 startAbilityForResult()方法显示 DeviceIDAbility 并获取返回的设备 ID,将返回的设备 ID 显示在主页面。主要代码如下:

```
Intent myintent = new Intent();
Operation operation = new Intent.OperationBuilder()
    .withBundleName("com.example.getonlinedevice")
    .withAbilityName("com.example.getonlinedevice.DeviceIDAbility")
    .build();
myintent.setOperation(operation);
startAbilityForResult(myintent, 100);
```

使用 startAbilityForResult()方法显示页面时,需要重写 onAbilityResult()方法接收 DeviceIDAbility 返回的数据,代码如下:

```
protected void onAbilityResult(int requestCode, int resultCode,
Intent resultIntent){
    switch (requestCode){
        case 100:
            switch (resultCode){
                case 101:
                //获取设备ID
                String deviceID = resultIntent.getStringParam("deviceID");
                TextField deviceMsg;
                deviceMsg = (TextField) findComponentById
                        (ResourceTable.Id_device_msg);
                //显示设备ID
                deviceMsg.setText("DeviceID:" + deviceID);
                break;
            }
        break;
    }
}
```

最后在 config.json 文件中注册这个 DeviceIDAbility，同时，需要添加一些与分布式相关的权限：

```
"reqPermissions": [
    {
        "name": "ohos.permission.GET_DISTRIBUTED_DEVICE_INFO"
    },
    {
        "name": "ohos.permission.ACCESS_DISTRIBUTED_ABILITY_GROUP"
    },
    {
        "name": "ohos.permission.DISTRIBUTED_DATASYNC"
    }
]
```

运行程序，进入 DeviceIDAbility 页面，可以看到如图 3-11 所示的设备列表，可以注意到，列表中只能获取到非本设备的其他设备信息。

3.5.3 根据设备 ID 调用 Page Ability

要实现跨设备迁移，需要实现 HarmonyOS 提供的 IAbilityContinuation 接口，该接口有以下 5 个方法。

(1) onStartContinuation()：Page 请求迁移后，系统首先回调此方法，开发者可以在此回调中决策当前是否可以执行迁移，例如，弹框让用户确认是否开始迁移。

(2) onSaveData()：如果 onStartContinuation() 返回 true，则系统回调此方法，开发者在此回调中保存必须传递到另外设备上以便恢复 Page 状态的数据。

(3) onRestoreData()：源设备上 Page 完成保存数据后，系统在目标设备上回调此方

图 3-11　获取可用设备信息效果图

法，开发者在此回调中接收用于恢复 Page 状态的数据。注意，在目标设备上的 Page 会重新启动其生命周期，无论其启动模式如何配置，系统回调此方法的时机在 onStart()之前。

（4）onCompleteContinuation()：目标设备上恢复数据一旦完成，系统就会在源设备上回调 Page 的此方法，以便通知应用迁移流程已结束。开发者可以在此检查迁移结果是否成功，并在此处理迁移结束的动作，例如，应用可以在迁移完成后终止自身生命周期。

（5）onRemoteTerminated()：若使用 continueAbilityReversibly()而不是 continueAbility()，则此后可以在源设备上使用 reverseContinueAbility()进行回迁操作。

请求迁移的设备需要调用以下方法发起迁移：

```
continueAbility(deviceId);
```

其中，deviceId 是目标设备 ID。调用该方法后，在源设备上会依次调用 onStartContinuation()方法和 onSaveData()方法，在目标设备会调用 onRestoreData()方法和 onCompleteContinuation()方法。其中 onSaveData()方法和 onRestoreData()方法用于在两个设备之间进行数据交互，两个方法都有一个类似于 Intent 类型的 IntentParams 类型参数，使用方法与 Intent 类似，通常是在 onRestoreData()中恢复源设备在 onSaveData()中保存的数据。

【例 3.5】　实现从一个设备迁移 Page Ability 到另一个设备，并在 TextField 中恢复迁移时的数据。首先创建名为 transfer_pageability.xml 的布局文件，包含了一个 Button 组件的一个 TextField 组件，代码如下：

```
<?xml version="1.0" encoding="utf-8"?>
<DirectionalLayout
    xmlns:ohos="http://schemas.huawei.com/res/ohos"
```

```xml
ohos:height="match_parent"
ohos:width="match_parent"
ohos:orientation="vertical">
<Button
    ohos:id="$+id:trans_button"
    ohos:height="50vp"
    ohos:width="match_parent"
    ohos:margin="10fp"
    ohos:text_alignment="center"
    ohos:background_element="$graphic:background_btn"
    ohos:text="迁移该 Page Ability"
    ohos:text_size="20fp"></Button>
<TextField
    ohos:id="$+id:textfiled"
    ohos:height="match_parent"
    ohos:width="match_parent"
    ohos:hint="请输入…"
    ohos:background_element="#f2f2f2"
    ohos:text_size="25vp"></TextField>
</DirectionalLayout>
```

然后创建一个名为 TransPageAbility 的 Page Ability 类，需要注意的是，该类必须要实现 IAbilityContinuation 接口并实现其所需的方法，主要代码如下：

```java
public class TransPageAbility extends Ability implements
IAbilityContinuation {
    private TextField textField;
    private Button transButton;
    private String data = "";
    @Override
    protected void onStart(Intent intent) {
        super.onStart(intent);
        super.setUIContent(ResourceTable.Layout_transfer_pageability);
        //申请权限
        requestPermission();
        transButton = (Button)findComponentById
        (ResourceTable.Id_trans_button);
        textField = (TextField)findComponentById
        (ResourceTable.Id_textfiled);
        transButton.setClickedListener(new Component.ClickedListener() {
            @Override
            public void onClick(Component component) {
                //显示可用设备页面
                Intent intent = new Intent();
                Operation operation = new Intent.OperationBuilder()
                .withBundleName("com.example.transdepartment")
                .withAbilityName("com.example.transdepartment
```

```java
                    .DeviceIDAbility").build();
                intent.setOperation(operation);
                startAbilityForResult(intent, 100);
            }
        });
        if(textField != null){
            textField.setText(data);
        }
    }
    protected void onAbilityResult(int requestCode, int resultCode,
                                    Intent resultIntent){
        switch (requestCode){
            case 100:
                switch (resultCode){
                    case 101:
                        //获取设备 ID
                        String deviceID = resultIntent.getStringParam
                                    ("deviceID");
                        try {
                            //迁移 Page Ability
                            continueAbility(deviceID);
                        }catch (IllegalStateException e){
                            e.printStackTrace();
                        }
                        break;
                }
            break;
        }
    }
    //请求授权
    private void requestPermission(){
        //在进行跨设备迁移之前,需要申请以下权限
        String[] permission = {"ohos.permission.DISTRIBUTED_DATASYNC"};
        List<String> applyPermissions = new ArrayList<>();
        for(String element : permission) {
            //验证是否已经获得该权限
            if(verifySelfPermission(element) != 0) {
                if(canRequestPermission(element)) {
                    //若未获得权限,将该权限加入权限列表
                    applyPermissions.add(element);
                }
                else {}
            }else{
            }
        }
        //申请权限
        requestPermissionsFromUser(applyPermissions.toArray
                        (new String[0]),0);
```

```java
    }
    //Page 请求迁移后,系统首先回调此方法
    @Override
    public boolean onStartContinuation() {
        return true;
    }
    //如果 onStartContinuation 返回 true,则系统回调此方法,
    //开发者在此回调中保存必须传递到另外设备上以便恢复 Page 状态的数据
@Override
public boolean onSaveData
(IntentParams intentParams) {
    intentParams.setParam("data", textField
    .getText());
    return true;
}
//源设备上 Page 完成保存数据后,系统在目标设备上回调
//此方法,开发者在回调中接收用于恢复 Page 状态的数据
@Override
public boolean onRestoreData
(IntentParams intentParams) {
    data = String.valueOf(intentParams.getParam
    ("data"));
    return true;
}
//目标设备上恢复数据一旦完成,系统就会在源设备上
//回调 Page 的此方法,以便通知应用迁移流程已结束
@Override
public void onCompleteContinuation(int i) {}
}
```

需要读者注意以下几点。

(1) onStartContinuation()方法、onSaveData()方法和 onRestoreData()方法都必须返回 true,如果读者使用 IDE 的自动生成代码功能,这几个方法都会默认返回 false,需要将它们的返回值改成 true。

(2) HarmonyOS 中有一些权限是敏感权限,需要在 Java 代码中进行动态注册,并不是在 config.json 文件中声明就可以了,本例需要允许不同设备间数据交换的权限,因此需要使用 Java 代码申请 ohos.permission.DISTRIBUTED_DATASYNC 权限,首次申请该权限时,系统会弹出如图 3-12 所示的对话框,需要用户的授权。

(3) 在 onRestoreData()方法中恢复迁移的数据时,一般只保存传递过来的数据,而不会直接在此方法中直接使用组件显示数据,因为 onRestoreData()方法在 onStart()方法之前调用,此时还未获取到组件对象,这些组件还都为 null,因此一般在 onStart()方法中获取组件对象后,再使用这些组件显示数据。

(4) 本例使用了 3.5.2 节实现的 DeviceIDAbility 获取可用设备列表,在进行跨设备迁移之前,会弹出一个可用设备列表供用户选择,需要用户选择一个设备,并返回该设备的 ID,然后在 onAbilityResult()方法中调用 continueAbility()方法迁移 Page Ability。

运行程序,确定授权,并在 TextField 组件中输入一些内容,最后单击"迁移该 Page Ability"按钮,跳转到当前可用设备信息列表页面,选择相应的设备后,会在目标设备中弹出该 Page Ability,并且在 TextFiled 组件中恢复源设备上 TextField 组件中的数据,如图 3-13 和图 3-14 所示。注意,若目标设备未安装该 App,则会先从应用市场安装该 App,只要目标设备安装了该 App,无论目标设备是否打开了该 App,都会自动弹出这个被迁移的 Page Ability。

图 3-12　申请权限对话框

图 3-13　源设备

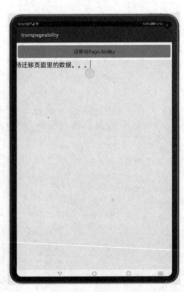
图 3-14　目标设备

3.6　AbilitySlice 间导航

作为使用 Page 模板的 Ability 的基本单位,AbilitySlice 为服务逻辑和 UI 显示提供了带有载体的功能。本节将介绍 AbilitySlice 的基础用法以及 AbilitySlice 间的导航方式。

3.6.1　AbilitySlice 的基础用法

AbilitySlice 只是普通的 Java 类,并不需要像 Page Ability 那样在 config.json 中添加注册信息,但必须继承 ohos.aafwk.ability.AbilitySlice 类,一个简单的 AbilitySlice 的代码如下:

```java
public class MyFirstSlice extends AbilitySlice {
    public void onStart(Intent intent){
        super.onStart(intent);
        super.setUIContent(ResourceTable.Layout_ability_main);
    }
}
```

Page Ability 可以直接加载布局文件，也可以导航到 AbilitySlice，由 AbilitySlice 负责装载布局文件。Page Ability 导航到 AbilitySlice 的代码如下：

```
public class MainAbility extends Ability {
    @Override
    public void onStart(Intent intent) {
        super.onStart(intent);
        super.setMainRoute(MyFirstSlice.class.getName());
    }
}
```

3.6.2 同一 Page 间导航

当发起导航的 AbilitySlice 和导航目标的 AbilitySlice 处于同一个 Page 时，可以通过 present()方法实现导航。如下代码展示使用 present()方法导航到 MyFirstSlice：

```
present(new MyFirstSlice(), new Intent());
```

其中，present()方法的两个参数：第一个表示要导航到的目的 AbilitySlice 对象，本例是 MyFirstSlice 对象，第二个参数是 Intent 对象，可以用来传递数据，与 Page Ability 中的 Intent 对象用法一致。

若要实现从 MyFirstSlice 返回数据，可以使用 presentForResult()方法，该方法的原型如下：

```
public final void presentForResult(AbilitySlice targetSlice, Intent intent,
    int requestCode);
```

其中，前两个参数与 present()方法含义相同，第三个参数用于标识导航到目标 AbilitySlice 的源 AbilitySlice。使用该方法导航到目的 AbilitySlice 时，还需在源 AbilitySlice 重写 onResult()方法，代码如下：

```
protected void onResult(int requestCode, Intent resultIntent){
    switch(requestCode){
        case 100:
            //处理返回的数据
            break;
    }
}
```

在目标 Ability 中，需要使用 terminate()方法关闭当前 AbilitySlice，在关闭前调用 setResult()方法即可返回数据，代码如下：

```
Intent intent = new Intent();
intent.setParam("data", 10);
setResult(intent);
terminate();
```

【例3.6】 实现同一个 Page Ability 中的不同 AbilitySlice 之间导航，由一个 AbilitySlice 导航到另一个 AbilitySlice，最后接收一个返回值。

首先创建一个名为 main_abilityslice.xml 的布局文件，该布局文件包含了一个 Button 组件，用于导航到 TargetSlice，一个 Text 组件用于显示返回的数据，代码如下：

```xml
<?xml version="1.0" encoding="utf-8"?>
<DirectionalLayout
    xmlns:ohos="http://schemas.huawei.com/res/ohos"
    ohos:height="match_parent"
    ohos:width="match_parent"
    ohos:alignment="center"
    ohos:orientation="vertical">
    <Button
        ohos:id="$+id:button"
        ohos:height="50vp"
        ohos:width="match_parent"
        ohos:background_element="$graphic:background_btn"
        ohos:margin="10fp"
        ohos:text="导航到 MyFirstSlice"
        ohos:text_size="20fp"></Button>
    <Text
        ohos:id="$+id:text"
        ohos:height="50vp"
        ohos:width="match_parent"
        ohos:text_alignment="center"
        ohos:text="待接收数据"
        ohos:text_size="20fp"></Text>
</DirectionalLayout>
```

然后创建一个名为 targetslice.xml 的布局文件，仅包含一个 Button 组件，用于关闭当前 AbilitySlice。代码如下：

```xml
<?xml version="1.0" encoding="utf-8"?>
<DirectionalLayout
    xmlns:ohos="http://schemas.huawei.com/res/ohos"
    ohos:height="match_parent"
    ohos:width="match_parent"
    ohos:orientation="vertical">
    <Button
        ohos:id="$+id:back"
        ohos:height="50fp"
        ohos:width="match_parent"
        ohos:background_element="$graphic:background_btn"
        ohos:margin="10fp"
        ohos:text="关闭 AbilitySlice"
        ohos:text_size="20vp"></Button>
</DirectionalLayout>
```

布局文件已准备完毕，现在编写 Java 代码，首先创建一个名为 MainAbilitySlice 的 AbilitySlice 类，代码如下：

```java
public class MainAbilitySlice extends AbilitySlice {
    private Button button;
    private Text text;
    @Override
    public void onStart(Intent intent) {
        super.onStart(intent);
        super.setUIContent(ResourceTable.Layout_main_abilityslice);
        button =(Button)findComponentById(ResourceTable.Id_button);
        text = (Text)findComponentById(ResourceTable.Id_text);
        button.setClickedListener(new Component.ClickedListener() {
            @Override
            public void onClick(Component component) {
                Intent intent = new Intent();
                //导航到 TargetSlice
                presentForResult(new TargetSlice(), intent, 100);
            }
        });
    }
    public void onResult(int requestCode, Intent resultIntent){
        switch(requestCode){
            case 100:
                //接收返回的数据
                String data = resultIntent.getStringParam("data");
                text.setText(data);
                break;
        }
    }
}
```

接下来再创建一个名为 TargetSlice 的 AbilitySlice 类，TargetSlice 和 MainAbilitySlice 在同一个 Page Ability（即 MainAbility）下，代码如下：

```java
public class TargetSlice extends AbilitySlice {
    private Button button;
    public void onStart(Intent intent){
        super.onStart(intent);
        super.setUIContent(ResourceTable.Layout_targetslice);
        button = (Button)findComponentById(ResourceTable.Id_back);
        button.setClickedListener(new Component.ClickedListener() {
            @Override
            public void onClick(Component component) {
                Intent intent = new Intent();
                intent.setParam("data", "这是 MyFirstSlice 返回的数据");
```

```
                //将数据返回
                setResult(intent);
                //关闭TargetSlice
                terminate();
            }
        });
    }
}
```

运行程序,单击"导航到 TargetSlice"按钮,导航到 TargetSlice 后,单击关闭按钮,页面返回到 MainAbilitySlice,并显示了返回的数据,由 Ability 的 label 可知,两个 AbilitySlice 在同一个 Page Ability 之下,即实现了同一 Page Ability 间导航。效果如图 3-15 所示。

图 3-15　同一 Page Ability 间导航示意图

3.6.3　不同 Page 间导航

AbilitySlice 作为 Page 的内部单元,以 Action 的形式对外暴露,因此可以通过配置 Intent 的 Action 导航到目标 AbilitySlice。因此,AbilitySlice 在不同的 Page 之间导航只需设置 action 即可,代码如下:

```
Intent intent = new Intent();
Operation operation = new Intent.OperationBuilder()
    .withBundleName("com.example.abilityslicedemo")
    .withAbilityName("com.example.abilityslicedemo.MainAbility")
    .withAction("action.test")
```

```
        .build();
intent.setOperation(operation);
startAbility(intent);
```

跳转到指定 Page Ability 的指定 AbilitySlice 时需要在指定的 Page Ability 中将 AbilitySlice 绑定一个 action 值，代码如下：

```
super.addActionRoute("action.test", TargetSlice.class.getName());
```

这样，在跳转页面时通过 withAction()方法指定 action 的值为"action.test"，便可跳转至 TargetSlice。

3.7 生命周期

3.7.1 Page Ability 的生命周期

生命周期是指 Page Ability 从创建、显示到销毁所经历的所有状态以及每个状态下的回调方法的总称。Page Ability 类提供许多回调方法，这些回调方法会让 Page Ability 知晓状态的变换：系统正在创建、显示一个 Page Ability，或者正在销毁一个 Page Ability。在生命周期不同回调方法中可以完成不同的工作，如在 onStart()方法中加载布局、初始化组件等操作。

Page Ability 生命周期的不同状态及其回调方法如图 3-16 所示。

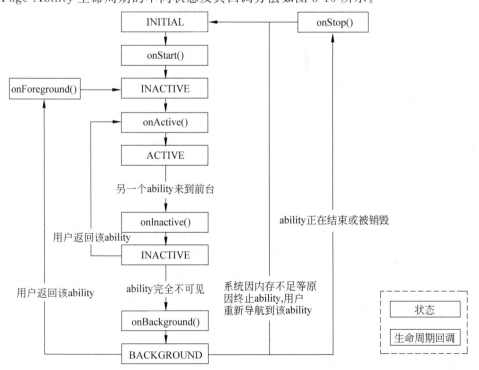

图 3-16　Page Ability 的生命周期示意图

Page Ability 一共有 6 个生命周期方法。

（1）onStart()：首次创建 Page 实例时，触发该回调。对于一个 Page 实例，该回调在其生命周期过程中仅触发一次，Page 在该逻辑后将进入 INACTIVE 状态。必须重写该方法，并在此配置默认展示的 AbilitySlice 或者加载一个布局文件。

（2）onActive()：Page 会在进入 INACTIVE 状态后来到前台，然后系统调用此回调。Page 在此之后进入 ACTIVE 状态，该状态是应用与用户交互的状态。Page 将保持在此状态，除非某类事件发生导致 Page 失去焦点，如用户单击返回键或导航到其他 Page。当此类事件发生时，会触发 Page 回到 INACTIVE 状态，系统将调用 onInactive() 回调。此后，Page 可能重新回到 ACTIVE 状态，系统将再次调用 onActive() 回调。因此，开发者通常需要成对实现 onActive() 和 onInactive()，并在 onActive() 中获取在 onInactive() 中被释放的资源。

（3）onInactive()：当 Page 失去焦点时，系统将调用此回调，此后 Page 进入 INACTIVE 状态。开发者可以在此回调中实现 Page 失去焦点时应表现的恰当行为。

（4）onBackground()：如果 Page 不再对用户可见，系统将调用此回调通知开发者用户进行相应的资源释放，此后 Page 进入 BACKGROUND 状态。开发者应该在此回调中释放 Page 不可见时无用的资源，或在此回调中执行较为耗时的状态保存操作。

（5）onForeground()：处于 BACKGROUND 状态的 Page 仍然驻留在内存中，当重新回到前台时（如用户重新导航到此 Page），系统将先调用 onForeground() 回调通知开发者，而后 Page 的生命周期状态回到 INACTIVE 状态。开发者应当在此回调中重新申请在 onBackground() 中释放的资源，最后 Page 的生命周期状态进一步回到 ACTIVE 状态，系统将通过 onActive() 回调通知开发者用户。

（6）onStop()：系统将要销毁 Page 时，会触发此回调函数，通知用户进行系统资源的释放。销毁 Page 的可能原因包括以下几个方面。

- 用户通过系统管理能力关闭指定 Page，如使用任务管理器关闭 Page。
- 用户行为触发 Page 的 terminateAbility() 方法调用，如使用应用的退出功能。
- 配置变更导致系统暂时销毁 Page 并重建。
- 系统出于资源管理目的，自动触发对处于 BACKGROUND 状态 Page 的销毁。

3.7.2 AbilitySlice 的生命周期

AbilitySlice 作为 Page 的组成单元，其生命周期是依托于其所属 Page 生命周期的。AbilitySlice 和 Page 具有相同的生命周期状态和同名的回调，当 Page 生命周期发生变化时，它的 AbilitySlice 也会发生相同的生命周期变化。此外，AbilitySlice 还具有独立于 Page 的生命周期变化，这发生在同一 Page 中的 AbilitySlice 之间导航时，此时 Page 的生命周期状态不会改变。

AbilitySlice 生命周期回调与 Page 的相应回调类似，因此不再赘述。由于 AbilitySlice 承载具体的页面，开发者必须重写 AbilitySlice 提供的 onStart() 回调方法，并在此方法中通过 setUIContent() 方法设置页面，代码如下：

```
protected void onStart(Intent intent) {
    super.onStart(intent);
    setUIContent(ResourceTable.Layout_main_layout);
}
```

  AbilitySlice 实例创建和管理通常由应用负责，系统仅在特定情况下会创建 AbilitySlice 实例。例如，通过导航启动某个 AbilitySlice 时，是由系统负责实例化；但是在同一 Page 的不同 AbilitySlice 间导航时则由应用负责实例化。

第4章 布　局

UI界面由很多个组件（Component）组成，为了将众多组件按用户的要求在屏幕上摆放，HarmonyOS提供了强大的布局支持，通过使用布局来设置各个组件之间的摆放关系（先后顺序、对齐方式、相对位置等），从而编写出精美的界面。本章主要介绍HarmonyOS的各种布局以及如何使用这些布局。

通过阅读本章，读者可以掌握：
- 方向布局。
- 依赖布局。
- 栈布局。
- 表格布局。
- 位置布局。
- 自适应盒子布局。

4.1　Java UI框架概述

移动应用是通过界面将数据呈现给用户并与用户完成交互的，任何一个应用的界面都是通过布局以及组件来设计实现的。

界面中的元素都称为组件，根据其功能和特点的不同，可将HarmonyOS的UI组件分为两类：一是Component，为用户提供内容显示，用户能与之交互。二是ComponentContainer，用于容纳其他Component和ComponentContainer对象的容器。本章主要介绍ComponentContainer，也就是布局，Component将在第5章详细介绍。

组件和布局根据一定的层级结构进行组合形成用户界面。单体组件在未被添加到容器组件中时，既无法显示也无法交互，因此一个用户界面至少包含一个容器组件。在布局中，可以添加组件，也可以添加其他布局，从而可以形成多种多样的UI样式，这便引出了组件树的概念，如图4-1所示。

布局把Component和ComponentContainer以树状的层级结构进行组织，这样的一个布局就称为组件树。组件树的特点是仅有一个根组件，其他组件有且仅有一个父节点，组件之间的关系受到父节点的规则约束。

在Java UI框架中，提供了两种编写布局的方式。

(1) 在代码中创建布局：使用Java代码创建Component和ComponentContainer对象，使用addComponent()方法将组件添加到布局中，最后使用setUIContent(ComponentContainer root)方法设置界面入口，并渲染到AbilitySlice中。

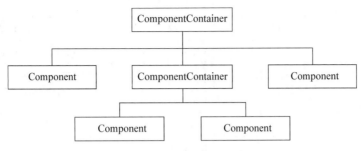

图 4-1　HarmonyOS 的组件树

（2）在 XML 文件中创建布局：XML 声明布局的方式更加简便直观，具体方法在第 3 章中已介绍过。

4.2　方向布局

方向布局（DirectionalLayout）是 Java UI 中的一种常用组件布局，用于指定内部组件的摆放顺序，将组件按照水平或者垂直方向排布，能够方便地对齐布局内的组件，如图 4-2 所示。

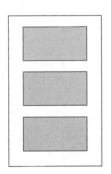

图 4-2　DirectionalLayout 示意图

4.2.1　支持的 XML 属性

DirectionalLayout 的自有 XML 属性如表 4-1 所示。

表 4-1　DirectionalLayout 的自有 XML 属性

属性名称	中文描述	取　　值	取值说明
alignment	对齐方式	left	表示左对齐
		top	表示顶部对齐
		right	表示右对齐
		bottom	表示底部对齐
		horizontal_center	表示水平居中对齐

续表

属性名称	中文描述	取值	取值说明
alignment	对齐方式	vertical_center	表示垂直居中对齐
		center	表示居中对齐
		start	表示靠起始端对齐
		end	表示靠结束端对齐
orientation	子布局排列方向	horizontal	表示水平方向布局
		vertical	表示垂直方向布局
total_weight	所有子视图的权重之和	float 类型	可以直接设置浮点数值,也可以引用 float 浮点数资源

DirectionalLayout 所包含组件可支持的 XML 属性如表 4-2 所示。

表 4-2　DirectionalLayout 所包含组件可支持的 XML 属性

属性名称	中文描述	取值	取值说明
layout_alignment	对齐方式	left	表示左对齐
		top	表示顶部对齐
		right	表示右对齐
		bottom	表示底部对齐
		horizontal_center	表示水平居中对齐
		vertical_center	表示垂直居中对齐
		center	表示居中对齐
weight	比重	float 类型	可以直接设置浮点数值,也可以引用 float 浮点数资源

4.2.2　排列方式

DirectionalLayout 的排列方向(orientation)分为水平(horizontal)或者垂直(vertical)方向。使用 orientation 设置布局内组件的排列方式,默认为垂直排列。

垂直排列示例代码如下:

```
<?xml version="1.0" encoding="utf-8"?>
<DirectionalLayout
    xmlns:ohos="http://schemas.huawei.com/res/ohos"
    ohos:width="match_parent"
    ohos:height="match_parent"
    ohos:background_element="#dddddd"
    ohos:alignment="vertical_center"
    ohos:orientation="vertical"
    ohos:padding="20vp">
```

```xml
    <Button
        ohos:width="match_parent"
        ohos:height="140vp"
        ohos:bottom_margin="10vp"
        ohos:background_element="#00FFFF"
        ohos:text="Button 1"
        ohos:text_size="30fp"/>
    <Button
        ohos:width="match_parent"
        ohos:height="140vp"
        ohos:bottom_margin="10vp"
        ohos:background_element="#00FFFF"
        ohos:text="Button 2"
        ohos:text_size="30fp"/>
    <Button
        ohos:width="match_parent"
        ohos:height="140vp"
        ohos:bottom_margin="10vp"
        ohos:background_element="#00FFFF"
        ohos:text="Button 3"
        ohos:text_size="30fp"/>
</DirectionalLayout>
```

水平排列示例代码如下：

```xml
<?xml version="1.0" encoding="utf-8"?>
<DirectionalLayout
    xmlns:ohos="http://schemas.huawei.com/res/ohos"
    ohos:width="match_parent"
    ohos:height="match_parent"
    ohos:background_element="#dddddd"
    ohos:alignment="vertical_center"
    ohos:orientation="horizontal"
    ohos:padding="20vp">
    <Button
        ohos:width="100vp"
        ohos:height="140vp"
        ohos:right_margin="10vp"
        ohos:background_element="#00FFFF"
        ohos:text="Button 4"
        ohos:text_size="20fp"/>
    <Button
        ohos:width="100vp"
        ohos:height="140vp"
        ohos:right_margin="10vp"
        ohos:background_element="#00FFFF"
        ohos:text="Button 5"
        ohos:text_size="20fp"/>
    <Button
        ohos:width="100vp"
```

```
            ohos:height="140vp"
            ohos:right_margin="10vp"
            ohos:background_element="#00FFFF"
            ohos:text="Button 6"
            ohos:text_size="20fp"/>
</DirectionalLayout>
```

如图 4-3 所示,图 4-3(a)中 Button 1、Button 2、Button 3 使用了垂直排列,图 4-3(b)中 Button 4、Button 5、Button 6 使用了水平排列。

(a) 垂直排列　　　　　　(b) 水平排列

图 4-3　排列方式效果示例

4.2.3　对齐方式

DirectionalLayout 使用 alignment 属性来控制所有内部组件的对齐方式。当某一组件需要单独设置对齐方式时,需要使用组件的 layout_alignment 属性来控制自身在布局中的对齐方式。对齐方式和排列方式密切相关,当排列方式为水平方向时,可选的对齐方式只有作用于垂直方向的类型(top、bottom、vertical_center、center),其他对齐方式不会生效。当排列方式为垂直方向时,可选的对齐方式只有作用于水平方向的类型(left、right、start、end、horizontal_center、center),其他对齐方式不会生效。

垂直排列对齐方式示例代码如下:

```
<?xml version="1.0" encoding="utf-8"?>
<DirectionalLayout
    xmlns:ohos="http://schemas.huawei.com/res/ohos"
    ohos:height="match_parent"
```

```xml
    ohos:width="match_parent"
    ohos:background_element="#dddddd"
    ohos:orientation="vertical"
    ohos:padding="32">
    <Button
        ohos:width="100vp"
        ohos:height="100vp"
        ohos:background_element="#00FFFF"
        ohos:layout_alignment="left"
        ohos:text="Button 1"
        ohos:text_size="20fp"/>
    <Button
        ohos:width="100vp"
        ohos:height="100vp"
        ohos:background_element="#00FFFF"
        ohos:layout_alignment="horizontal_center"
        ohos:text="Button 2"
        ohos:text_size="20fp"/>
    <Button
        ohos:width="100vp"
        ohos:height="100vp"
        ohos:background_element="#00FFFF"
        ohos:layout_alignment="right"
        ohos:text="Button 3"
        ohos:text_size="20fp"/>
</DirectionalLayout>
```

水平排列对齐方式示例代码如下：

```xml
<?xml version="1.0" encoding="utf-8"?>
<DirectionalLayout
    xmlns:ohos="http://schemas.huawei.com/res/ohos"
    ohos:height="match_parent"
    ohos:width="match_parent"
    ohos:background_element="#dddddd"
    ohos:orientation="horizontal"
    ohos:padding="32">
    <Button
        ohos:width="100vp"
        ohos:height="100vp"
        ohos:background_element="#00FFFF"
        ohos:layout_alignment="top"
        ohos:text="Button 4"
        ohos:text_size="20fp"/>
    <Button
        ohos:width="100vp"
        ohos:height="100vp"
        ohos:background_element="#00FFFF"
        ohos:layout_alignment="vertical_center"
```

```
            ohos:text="Button 5"
            ohos:text_size="20fp"/>
    <Button
            ohos:width="100vp"
            ohos:height="100vp"
            ohos:background_element="#00FFFF"
            ohos:layout_alignment="bottom"
            ohos:text="Button 6"
            ohos:text_size="20fp"/>
</DirectionalLayout>
```

如图 4-4 所示，垂直排列的 Button 1、Button 2、Button 3 分别使用 left、horizontal_center、right 三种对齐方式，水平排列的 Button 4、Button 5、Button 6 分别使用 top、vertical_center、bottom 三种对齐方式。

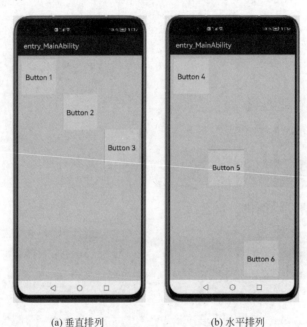

(a) 垂直排列　　　　　　　(b) 水平排列

图 4-4　不同对齐方式的效果示例

4.2.4　权重

权重(weight)是一种百分比分配大小的布局方式，父布局可分配宽度(高度)与组件宽度(高度)的计算方法如下所示：

父布局可分配宽度(高度)=父布局宽度(高度)-所有子组件宽度(高度)之和。

组件宽度(高度)=组件权重÷所有组件权重之和×父布局可分配宽度(高度)。

权重分配示例代码如下：

```
<?xml version="1.0" encoding="utf-8"?>
<DirectionalLayout
```

```
    xmlns:ohos="http://schemas.huawei.com/res/ohos"
    ohos:width="match_parent"
    ohos:height="match_parent"
    ohos:background_element="#dddddd"
    ohos:orientation="vertical">
    <Button
        ohos:width="200vp"
        ohos:height="0vp"
        ohos:weight="2"
        ohos:background_element="#00FFFF"
        ohos:text="Button 1"
        ohos:text_size="30fp"/>
    <Button
        ohos:width="200vp"
        ohos:height="0vp"
        ohos:weight="1"
        ohos:background_element="#ffffff"
        ohos:text="Button 2"
        ohos:text_size="30fp"/>
    <Button
        ohos:width="200vp"
        ohos:height="0vp"
        ohos:weight="1"
        ohos:background_element="#00FFFF"
        ohos:text="Button 3"
        ohos:text_size="30fp"/>
</DirectionalLayout>
```

水平排列的代码与垂直排列类似,这里不再给出,效果如图4-5所示。

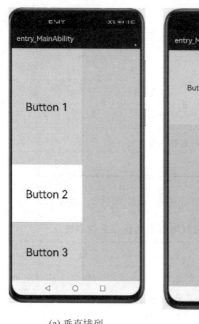

(a) 垂直排列　　　　　　　(b) 水平排列

图4-5　权重分配组件大小效果示例

4.3 依赖布局

依赖布局（DependentLayout）也是 Java UI 中的一种常见布局。如图 4-6 所示，与 DirectionalLayout 整齐单一的排布方式不同，DependentLayout 的排布方式更加多样化，每个组件可以指定相对于其他同级元素的位置，或者指定相对于父组件的位置。

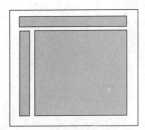

图 4-6 DependentLayout 示意图

4.3.1 支持的 XML 属性

DependentLayout 的自有 XML 属性如表 4-3 所示。

表 4-3 DependentLayout 的自有 XML 属性

属性名称	中文描述	取值	取值说明
alignment	对齐方式	left	表示左对齐
		top	表示顶部对齐
		right	表示右对齐
		bottom	表示底部对齐
		horizontal_center	表示水平居中对齐
		vertical_center	表示垂直居中对齐
		center	表示居中对齐
		start	表示靠起始端对齐
		end	表示靠结束端对齐

DependentLayout 控制同级别组件位置的属性如表 4-4 所示。

表 4-4 控制同级别组件位置的属性

布局属性	描述
above	将下边缘与另一个子组件的上边缘对齐
below	将上边缘与另一个子组件的下边缘对齐
start_of	将结束边与另一个子组件的起始边对齐
end_of	将起始边与另一个子组件的结束边对齐

续表

布 局 属 性	描　　述
left_of	将右边缘与另一个子组件的左边缘对齐
right_of	将左边缘与另一个子组件的右边缘对齐
align_baseline	将子组件的基线与另一个子组件的基线对齐
align_left	将左边缘与另一个子组件的左边缘对齐
align_top	将上边缘与另一个子组件的上边缘对齐
align_right	将右边缘与另一个子组件的右边缘对齐
align_bottom	将底边与另一个子组件的底边对齐
align_start	将起始边与另一个子组件的起始边对齐
align_end	将结束边与另一个子组件的结束边对齐

表 4-4 中的 6 个属性需要指定组件的 ID。

DependentLayout 控制父容器和子组件位置的属性如表 4-5 所示。

表 4-5　控制父容器和子组件位置的属性

布 局 属 性	描　　述
align_parent_left	将左边缘与父组件的左边缘对齐
align_parent_top	将上边缘与父组件的上边缘对齐
align_parent_right	将右边缘与父组件的右边缘对齐
align_parent_bottom	将底边与父组件的底边对齐
align_parent_start	将起始边与父组件的起始边对齐
align_parent_end	将结束边与父组件的结束边对齐
center_in_parent	将子组件保持在父组件的中心
horizontal_center	将子组件保持在父组件水平方向的中心
vertical_center	将子组件保持在父组件垂直方向的中心

表 4-5 中的属性都是布尔类型，只能设置为 true 或 false。

4.3.2　排列方式

DependentLayout 的排列方式是相对于其他同级组件或者父组件的位置进行布局的。

1. 相对于同级组件的对齐

对于同级组件来说，有两种不同的对齐方式：根据位置对齐和根据边对齐。

above、below、start_of、end_of、left_of 和 right_of 均是根据位置对齐的方式。简单理解就是组件在同级基准组件的相应位置。如 above 对齐方式，将下边缘与同级基准组件的上边缘对齐，即位于同级组件的上侧。其余几种对齐方式遵循的逻辑与此类似，需要注意的是 start_of 和 end_of 会跟随当前布局起始方向变化。

位置对齐示例代码如下：

```xml
<?xml version="1.0" encoding="utf-8"?>
<DependentLayout
    xmlns:ohos="http://schemas.huawei.com/res/ohos"
    ohos:height="match_parent"
    ohos:width="match_parent"
    ohos:background_element="#dddddd">
    <Text
        ohos:id="$+id:text_start"
        ohos:height="200vp"
        ohos:width="100vp"
        ohos:start_of="$id:text_center"
        ohos:background_element="#FF9912"
        ohos:padding="8vp"
        ohos:text="start_of"
        ohos:text_size="20fp"/>
    <Text
        ohos:id="$+id:text_end"
        ohos:height="200vp"
        ohos:width="100vp"
        ohos:end_of="$id:text_center"
        ohos:background_element="#FF9912"
        ohos:padding="8vp"
        ohos:text="end_of"
        ohos:text_size="20fp"/>
    <Text
        ohos:id="$+id:text_center"
        ohos:height="100vp"
        ohos:width="100vp"
        ohos:background_element="#00FFFF"
        ohos:center_in_parent="true"
        ohos:text="center"
        ohos:text_alignment="center"
        ohos:text_size="20fp"/>
    <Text
        ohos:id="$+id:text_left"
        ohos:height="match_content"
        ohos:width="match_content"
        ohos:left_of="$id:text_center"
        ohos:background_element="#00FFFF"
        ohos:padding="8vp"
        ohos:text="left_of"
        ohos:text_size="20fp"/>
    <Text
        ohos:id="$+id:text_right"
        ohos:height="match_content"
        ohos:width="match_content"
        ohos:right_of="$id:text_center"
        ohos:background_element="#00FFFF"
        ohos:padding="8vp"
        ohos:text="right_of"
        ohos:text_size="18fp"/>
```

```xml
<Text
    ohos:id="$+id:text_above"
    ohos:height="match_content"
    ohos:width="match_content"
    ohos:above="$id:text_center"
    ohos:background_element="#00FFFF"
    ohos:padding="8vp"
    ohos:text="above"
    ohos:text_size="20fp"/>
<Text
    ohos:id="$+id:text_below"
    ohos:height="match_content"
    ohos:width="match_content"
    ohos:below="$id:text_center"
    ohos:background_element="#00FFFF"
    ohos:padding="8vp"
    ohos:text="below"
    ohos:text_size="20fp"/>
</DependentLayout>
```

位置对齐的几种不同对齐方式的效果如图 4-7 所示。

align_top、align_bottom、align_left、align_right、align_start 和 align_end 均是根据边对齐的方式。简单理解就是组件与同级基准组件对应边对齐。如 align_top 对齐方式，即将当前组件与同级组件的上边缘对齐。其他几种对齐方式遵循的逻辑与此类似，需要注意的是 align_start 和 align_end 会跟随当前布局起始方向变化。

边对齐的代码与位置对齐类似，这里不再给出，效果如图 4-8 所示（注此处的 center 为基准）。

图 4-7　位置对齐的效果示例　　　　图 4-8　边对齐的效果示例

2. 相对于父级组件的对齐

以父级组件为基准设置内部组件的位置,若没有父级组件,则以最外侧的布局组件为基准。示例代码如下:

```xml
<?xml version="1.0" encoding="utf-8"?>
<DependentLayout
    xmlns:ohos="http://schemas.huawei.com/res/ohos"
    ohos:height="match_parent"
    ohos:width="match_parent"
    ohos:background_element="#dddddd">
    <Text
        ohos:height="250vp"
        ohos:width="110vp"
        ohos:align_parent_left="true"
        ohos:background_element="#FF9912"
        ohos:padding="12vp"
        ohos:multiple_lines="true"
        ohos:text="align_parent_left"
        ohos:text_size="20fp"
        ohos:text_color="#FFFFFF"/>
    <Text
        ohos:height="100vp"
        ohos:width="100vp"
        ohos:align_parent_right="true"
        ohos:background_element="#00FFFF"
        ohos:padding="8vp"
        ohos:multiple_lines="true"
        ohos:text="align_parent_right"
        ohos:text_size="20fp"/>
    <Text
        ohos:height="100vp"
        ohos:width="100vp"
        ohos:align_parent_top="true"
        ohos:background_element="#00FFFF"
        ohos:padding="8vp"
        ohos:multiple_lines="true"
        ohos:text="align_parent_top"
        ohos:text_size="20fp"/>
    <Text
        ohos:height="100vp"
        ohos:width="100vp"
        ohos:align_parent_bottom="true"
        ohos:background_element="#00FFFF"
        ohos:padding="8vp"
        ohos:multiple_lines="true"
        ohos:text="align_parent_bottom"
        ohos:text_size="20fp"/>
    <Text
```

```xml
        ohos:height="100vp"
        ohos:width="100vp"
        ohos:center_in_parent="true"
        ohos:background_element="#00FFFF"
        ohos:padding="8vp"
        ohos:multiple_lines="true"
        ohos:text="center_in_parent"
        ohos:text_size="20fp"/>
    <Text
        ohos:height="100vp"
        ohos:width="100vp"
        ohos:horizontal_center="true"
        ohos:background_element="#00FFFF"
        ohos:padding="8vp"
        ohos:multiple_lines="true"
        ohos:text="horizontal_center"
        ohos:text_size="20fp"/>
    <Text
        ohos:height="100vp"
        ohos:width="100vp"
        ohos:vertical_center="true"
        ohos:background_element="#00FFFF"
        ohos:padding="8vp"
        ohos:multiple_lines="true"
        ohos:text="vertical_center"
        ohos:text_size="20fp"/>
</DependentLayout>
```

相对于父级组件的几种不同对齐方式的效果如图 4-9 所示。

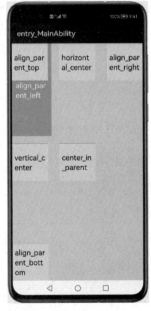

图 4-9 相对于父级组件对齐方式的效果示例

4.4 表格布局

表格布局(TableLayout)也是 Java UI 中的一种常见布局。能将组件按照行列的方式放在表格中,形成整洁有序的布局方式,如图 4-10 所示。

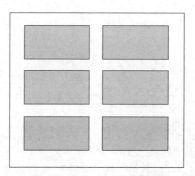

图 4-10　TableLayout 示意图

4.4.1 支持的 XML 属性

TableLayout 的自有 XML 属性如表 4-6 所示。

表 4-6　TableLayout 的自有 XML 属性

属性名称	中文描述	取值	取值说明
alignment_type	对齐方式	align_edges	表示 TableLayout 内的组件按边界对齐
		align_contents	表示 TableLayout 内的组件按边距对齐
column_count	列数	integer 类型	可以直接设置整型数值,也可以引用 integer 资源
row_count	行数	integer 类型	可以直接设置整型数值,也可以引用 integer 资源
orientation	排列方向	horizontal	表示水平方向布局。
		vertical	表示垂直方向布局。

4.4.2 设置行列数

若不设置行列数,则默认以一行多列的方式排列。

```
<?xml version="1.0" encoding="utf-8"?>
<TableLayout
    xmlns:ohos="http://schemas.huawei.com/res/ohos"
    ohos:height="match_parent"
    ohos:width="match_parent"
    ohos:background_element="#dddddd"
    ohos:padding="8vp"
    ohos:row_count="2"
    ohos:column_count="2">
    <Text
        ohos:height="100vp"
```

```xml
        ohos:width="100vp"
        ohos:background_element="#00FFFF"
        ohos:margin="8vp"
        ohos:text="1"
        ohos:text_alignment="center"
        ohos:text_size="50fp"/>
    <Text
        ohos:height="100vp"
        ohos:width="100vp"
        ohos:background_element="#00FFFF"
        ohos:margin="8vp"
        ohos:text="2"
        ohos:text_alignment="center"
        ohos:text_size="50fp"/>
    <Text
        ohos:height="100vp"
        ohos:width="100vp"
        ohos:background_element="#00FFFF"
        ohos:margin="8vp"
        ohos:text="3"
        ohos:text_alignment="center"
        ohos:text_size="50fp"/>
    <Text
        ohos:height="100vp"
        ohos:width="100vp"
        ohos:background_element="#00FFFF"
        ohos:margin="8vp"
        ohos:text="4"
        ohos:text_alignment="center"
        ohos:text_size="50fp"/>
</TableLayout>
```

如图 4-11 所示，图 4-11(a)为不设置行列数，图 4-11(b)为设置行列数为 2。

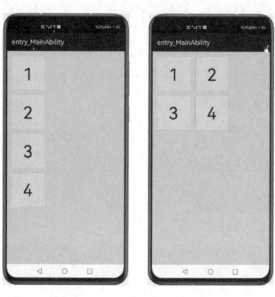

(a) 设置行列前　　　　　　(b) 设置行列后

图 4-11　设置 TableLayout 的行为 2，列为 2 效果示例

4.4.3 设置布局排列方向

布局排列方向默认为 horizontal，效果如图 4-11 所示，组件的添加顺序即组件上的数字。设置布局排列方向为 vertical，则效果如图 4-12（左）所示。如图 4-12（中）和图 4-12（右）分别为添加个数不足或是超过设定个数（4 个）时的效果。

图 4-12　设置布局排列方向为 vertical 的效果示例

4.4.4 设置对齐方式

TableLayout 通过 alignment_type 设置对齐方式，它提供两种对齐方式：边距对齐"align_contents"和边界对齐"align_edges"，默认为边距对齐"align_contents"。

```xml
<?xml version="1.0" encoding="utf-8"?>
<TableLayout
    xmlns:ohos="http://schemas.huawei.com/res/ohos"
    ohos:height="match_parent"
    ohos:width="match_parent"
    ohos:alignment_type="align_contents"
    ohos:background_element="#dddddd"
    ohos:row_count="2"
    ohos:column_count="2">
    <Text
        ohos:height="100vp"
        ohos:width="100vp"
        ohos:background_element="#00FFFF"
        ohos:top_margin="8vp"
        ohos:bottom_margin="8vp"
        ohos:right_margin="8vp"
        ohos:left_margin="108vp"
        ohos:padding="8vp"
        ohos:text="1"
        ohos:text_alignment="center"
        ohos:text_size="50fp"/>
```

```xml
    <Text
        ohos:height="100vp"
        ohos:width="100vp"
        ohos:background_element="#00FFFF"
        ohos:margin="8vp"
        ohos:padding="8vp"
        ohos:text="2"
        ohos:text_alignment="center"
        ohos:text_size="50fp"/>
    <Text
        ohos:height="100vp"
        ohos:width="100vp"
        ohos:background_element="#00FFFF"
        ohos:margin="8vp"
        ohos:padding="8vp"
        ohos:text="3"
        ohos:text_alignment="center"
        ohos:text_size="50fp"/>
    <Text
        ohos:height="100vp"
        ohos:width="100vp"
        ohos:background_element="#00FFFF"
        ohos:margin="8vp"
        ohos:padding="8vp"
        ohos:text="4"
        ohos:text_alignment="center"
        ohos:text_size="50fp"/>
</TableLayout>
```

如图 4-13 所示，分别是两种不同对齐方式的效果图。其中粗框为该 text 组件的范围，包括组件的边距（margin 值）。边距对齐在对齐时考虑组件边距，而边界对齐在对齐时只考虑组件自身的边界。

(a) 边距对齐效果示例　　(b) 边界对齐效果示例

图 4-13　边距对齐的效果示例

4.5 栈布局

栈布局(StackLayout)用来实现组件之间的层叠布局，StackLayout 直接在屏幕上开辟出一块空白的区域，添加到这个布局中的视图都是以层叠的方式显示，而它会把这些视图默认放到这块区域的左上角，第一个添加到布局中的视图显示在最底层，之后的组件会被添加到上层，依次堆叠，如图 4-14 所示。

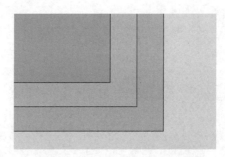

图 4-14　StackLayout 示意图

4.5.1　支持的 XML 属性

StackLayout 所包含组件可支持的 XML 属性如表 4-7 所示。

表 4-7　StackLayout 所包含组件可支持的 XML 属性

属性名称	中文描述	取值	取值说明
layout_alignment	对齐方式	left	表示左对齐
		top	表示顶部对齐
		right	表示右对齐
		bottom	表示底部对齐
		horizontal_center	表示水平居中对齐
		vertical_center	表示垂直居中对齐
		center	表示居中对齐

4.5.2　使用默认布局添加组件

StackLayout 内的组件默认放在区域的左上角，之后添加的组件默认放置在原先的上层。页面效果如图 4-15 所示。

```
<?xml version="1.0" encoding="utf-8"?>
<StackLayout
    xmlns:ohos="http://schemas.huawei.com/res/ohos"
    ohos:id="$+id:stack_layout"
    ohos:height="match_parent"
```

```
    ohos:width="match_parent">
    <Text
        ohos:text_alignment="bottom|horizontal_center"
        ohos:text_size="20fp"
        ohos:text="第一层"
        ohos:height="400vp"
        ohos:width="400vp"
        ohos:background_element="#3F56EA" />
    <Text
        ohos:text_alignment="bottom|horizontal_center"
        ohos:text_size="20fp"
        ohos:text="第二层"
        ohos:height="200vp"
        ohos:width="200vp"
        ohos:background_element="#00FFFF" />
    <Text
        ohos:text_alignment="center"
        ohos:text_size="20fp"
        ohos:text="第三层"
        ohos:height="100vp"
        ohos:width="100vp"
        ohos:background_element="#3F56EA" />
</StackLayout>
```

4.5.3 使用相对位置添加组件

StackLayout 使用 layout_alignment 属性指定组件的相对位置，如下使用该属性将组件位置从左上角移至右下角。效果如图 4-16 所示。

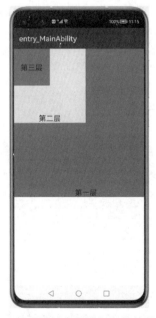 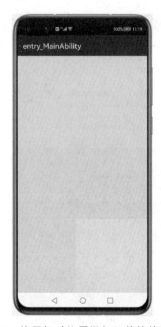

图 4-15　使用默认布局添加组件的效果示例　　图 4-16　使用相对位置添加组件的效果示例

```xml
<?xml version="1.0" encoding="utf-8"?>
<StackLayout
    xmlns:ohos="http://schemas.huawei.com/res/ohos"
    ohos:id="$+id:stack_layout"
    ohos:height="match_parent"
    ohos:width="match_parent"
    ohos:background_element="#dddddd">
    <Text
        ohos:height="200vp"
        ohos:width="200vp"
        ohos:layout_alignment="bottom|right"
        ohos:background_element="#00FFFF"/>
</StackLayout>
```

4.6 位置布局

位置布局(PositionLayout)用于为组件设定特殊的位置,如图 4-17 所示。在一些特殊场景下,如果需要使用代码控制组件的位置或者需要固定组件的位置,通常需要使用位置布局。在 PositionLayout 中,通过指定组件的(x,y)坐标值在屏幕上显示。起始坐标$(0,0)$默认在左上角。

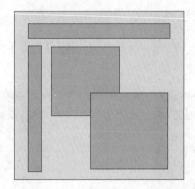

图 4-17　PositionLayout 示意图

```xml
<?xml version="1.0" encoding="utf-8"?>
<PositionLayout
    xmlns:ohos="http://schemas.huawei.com/res/ohos"
    ohos:id="$+id:position"
    ohos:height="match_parent"
    ohos:width="match_parent"
    ohos:background_element="#dddddd">
    <Text
        ohos:id="$+id:position_text_1"
        ohos:height="100vp"
        ohos:width="100vp"
        ohos:background_element="#9987CEFA"
        ohos:position_x="0vp"
```

```
            ohos:position_y="0vp"
            ohos:text="Text 1"
            ohos:text_alignment="center"
            ohos:text_size="20fp"/>
        <Text
            ohos:id="$+id:position_text_2"
            ohos:height="200vp"
            ohos:width="200vp"
            ohos:background_element="#9987CEFA"
            ohos:position_x="150vp"
            ohos:position_y="120vp"
            ohos:text="Text 2"
            ohos:text_alignment="center"
            ohos:text_size="20fp"/>
        <Text
            ohos:id="$+id:position_text_3"
            ohos:height="200vp"
            ohos:width="200vp"
            ohos:background_element="#9987CEFA"
            ohos:position_x="80vp"
            ohos:position_y="250vp"
            ohos:text="Text 3"
            ohos:text_alignment="center"
            ohos:text_size="20fp"/>
</PositionLayout>
```

PositionLayout 以坐标的形式控制组件的显示位置，允许组件相互重叠。

在 layout 目录下的 XML 文件中创建 PositionLayout 并添加多个组件，并通过 position_x 和 position_y 属性设置子组件的坐标。

如图 4-18 展示了使用 PositionLayout 布局的效果。其中 text1 组件左上角位于(0,0)的位置，也就是页面坐标的起始点。text2 与 text3 组件左上角分别位于(150, 120)与(80,250)，并且有重叠。

除了在 XML 文件中设置 position_x 和 position_y 属性的方法外，还可以在对应的 AbilitySlice 中通过 setPosition(int x, int y) 接口设置组件的位置。使用 AttrHelper 类的 vp2px() 方法转换单位 px 为 vp。代码如下：

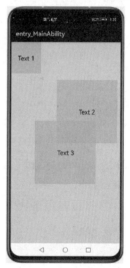

图 4-18　使用 PositionLayout 布局的效果示例

```
Text text1 = (Text)findComponentById(ResourceTable.Id_position_text_1);
Text text2 = (Text)findComponentById(ResourceTable.Id_position_text_2);
Text text3 = (Text)findComponentById(ResourceTable.Id_position_text_3);
text1.setPosition(AttrHelper.vp2px(0,AttrHelper.getDensity(this)),
AttrHelper.vp2px(0,AttrHelper.getDensity(this)));
text2.setPosition(AttrHelper.vp2px(150,AttrHelper.getDensity(this)),
```

```
AttrHelper.vp2px(120,AttrHelper.getDensity(this)));
text3.setPosition(AttrHelper.vp2px(80,AttrHelper.getDensity(this)),
AttrHelper.vp2px(250,AttrHelper.getDensity(this)));
```

4.7 自适应盒子布局

自适应盒子布局(AdaptiveBoxLayout)使得页面上的多个组件在不同屏幕尺寸设备上能够自适应调整列数,如图 4-19 所示。AdaptiveBoxLayout 主要有以下 4 个特点。

(1) 该布局中的每个子组件都用一个单独的"盒子"装起来,子组件设置的布局参数都是以盒子作为父布局生效,不以整个自适应布局为生效范围。

(2) 该布局中每个盒子的宽度固定为布局总宽度除以自适应得到的列数,高度为 match_content,每一行中的所有盒子按高度最高的进行对齐。

(3) 该布局水平方向是自动分块,因此水平方向不支持 match_content,布局水平宽度仅支持 match_parent 或固定宽度。

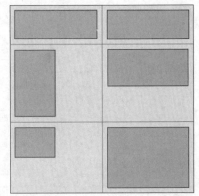

图 4-19 AdaptiveBoxLayout 示意图

(4) 自适应仅在水平方向进行了自动分块,纵向没有限制,因此如果某个子组件的高设置为 match_parent 类型,可能导致后续行无法显示。

4.7.1 常用方法

AdaptiveBoxLayout 布局常用方法如下。

(1) 添加一个自适应盒子布局规则:addAdaptiveRule(int minWidth, int maxWidth, int columns)。

(2) 移除一个自适应盒子布局规则:removeAdaptiveRule(int minWidth, int maxWidth, int columns)。

(3) 移除所有自适应盒子布局规则:clearAdaptiveRules()。

4.7.2 场景示例

利用下列规则实现对不同屏幕宽度的自适应盒子布局。

(1) addAdaptiveRule(0,1000,2):当布局宽度为 0~1000 时,组件每行显示 2 个。

(2) addAdaptiveRule(1000,2000,3):当布局宽度为 1000~2000 时,组件每行显示 3 个。

(3) addAdaptiveRule(2000,Integer MAX_VALVE,4):当布局宽度大于 2000 时,组件每行显示 4 个。

图 4-20(a)为添加规则后的效果图,按照盒子布局排列;图 4-20(b)为清除规则后的效果图,按照线性布局排列。当手机竖屏时,布局的宽度为 2340(2340>2000),故自适应盒子布局每行的组件个数为 3。

第 4 章 布局

(a) 添加规则后　　　　　(b) 清除规则后

图 4-20　添加与清除规则效果示例

如图 4-21 所示，当手机横屏时，布局的宽度为 2340(2340＞2000)，故自适应盒子布局每行的组件个数为 4。

图 4-21　手机横屏后的效果示例

```
<?xml version="1.0" encoding="utf-8"?>
<DirectionalLayout
    xmlns:ohos="http://schemas.huawei.com/res/ohos"
    ohos:height="match_parent"
    ohos:width="match_parent"
    ohos:background_element="#dddddd"
    ohos:orientation="vertical">
    <AdaptiveBoxLayout
        xmlns:ohos="http://schemas.huawei.com/res/ohos"
        ohos:height="0vp"
        ohos:width="match_parent"
        ohos:weight="2"
```

```xml
        ohos:id="$+id:adaptive_box_layout">
    <Text
        ohos:height="match_content"
        ohos:width="match_parent"
        ohos:background_element="#00FFFF"
        ohos:margin="10vp"
        ohos:padding="10vp"
        ohos:text="1"
        ohos:text_size="30fp" />
    <Text
        ohos:height="match_content"
        ohos:width="match_parent"
        ohos:background_element="#00FFFF"
        ohos:margin="10vp"
        ohos:padding="10vp"
        ohos:text="2"
        ohos:text_size="30fp" />
    <Text
        ohos:height="match_content"
        ohos:width="match_parent"
        ohos:background_element="#00FFFF"
        ohos:margin="10vp"
        ohos:padding="10vp"
        ohos:multiple_lines="true"
        ohos:text="3"
        ohos:text_size="30fp" />
    <Text
        ohos:height="match_content"
        ohos:width="match_parent"
        ohos:background_element="#00FFFF"
        ohos:margin="10vp"
        ohos:padding="10vp"
        ohos:text="4"
        ohos:text_size="30fp" />
</AdaptiveBoxLayout>
<DirectionalLayout
    ohos:height="0vp"
    ohos:width="match_parent"
    ohos:weight="1"
    ohos:orientation="horizontal"
    ohos:padding="10vp">
    <Button
        ohos:id="$+id:add_btn"
        ohos:padding="10vp"
        ohos:margin="10vp"
        ohos:background_element="#00FFFF"
        ohos:height="match_content"
```

```xml
            ohos:width="0vp"
            ohos:weight="1"
            ohos:text_size="22fp"
            ohos:text="addRules"/>
        <Button
            ohos:id="$+id:clear_btn"
            ohos:padding="10vp"
            ohos:margin="10vp"
            ohos:background_element="#00FFFF"
            ohos:height="match_content"
            ohos:width="0vp"
            ohos:weight="1"
            ohos:text_size="22fp"
            ohos:text="clearRules"/>
    </DirectionalLayout>
</DirectionalLayout>
```

Java 关键代码如下：

```java
//获取自适应布局的对象
AdaptiveBoxLayout adaptiveBoxLayout = (AdaptiveBoxLayout)findComponentById
                    (ResourceTable.Id_adaptive_box_layout);
//增加规则按钮
findComponentById(ResourceTable.Id_add_btn)
                    .setClickedListener((component-> {
    //增加规则 当布局宽度为 0-1000 时,组件每行显示 2 个
    adaptiveBoxLayout.addAdaptiveRule(0,1000,2);
    //增加规则 当布局宽度为 1000-2000 时,组件每行显示 3 个
    adaptiveBoxLayout.addAdaptiveRule(1000,2000,3);
    //增加规则 当布局宽度大于 2000 时,组件每行显示 4 个
    adaptiveBoxLayout.addAdaptiveRule(2000,Integer.MAX_VALUE,4);
    //更新布局
    adaptiveBoxLayout.postLayout();
}));
//清除规则按钮
findComponentById(ResourceTable.Id_clear_btn)
                    .setClickedListener((component-> {
    //移除所有规则
    adaptiveBoxLayout.clearAdaptiveRules();
    //更新布局
    adaptiveBoxLayout.postLayout();
}));
```

第 5 章 UI 组 件

本章主要讲解常见 UI 组件的使用方法及其特性。第 4 章中介绍了 HarmonyOS 的常用 UI 布局,而一个界面除了 UI 布局,组件也是非常重要的组成部分。组件是构建页面的核心,每个组件通过对数据和方法的简单封装,实现独立的可视、可交互功能单元。组件之间相互独立,随取随用,也可以在需求相同的地方重复使用。HarmonyOS 的 UI 常见组件可以分为三大类:显示组件、交互组件和高级组件。组件的具体使用场景,需要根据业务需求来选择使用。

通过阅读本章,读者可以掌握:
➢ 如何创建使用各类组件。
➢ 了解各类组件所支持的属性。
➢ 如何设置组件样式。

5.1 展 示 组 件

5.1.1 文本组件

文本组件(Text)是最常用的组件之一。Text 是用来显示字符串的组件,在界面上显示为一块文本区域。Text 作为一个基本组件有很多扩展,常见的有按钮组件 Button,文本编辑组件 TextField。Text 是 Component 类的子类之一,所以它能够使用 Component 类的所有公开的属性和方法,Text 类自身也提供了一些特殊的属性、方法、内部类和接口。

1. 创建 Text

(1) 布局文件中创建 Text。

在 layout 文件下创建 text.xml,在 XML 文件中声明布局和组件。例如,创建一个高为 150vp,长为 300vp 的 text 组件,组件具体属性代码如下:

```
<Text
    ohos:id="$+id:text" //当前组件唯一标识,在单个布局文件中不允许 id 标识重复
    ohos:width="300vp"
    ohos:height="150vp"
    ohos:text="Text 组件"     //显示文本
    ohos:background_element="$graphic:background_text"   //设置 Text 背景
    ohos:text_size="66fp"
    ohos:text_color="white"
    ohos:italic="true"
```

```
    ohos:text_weight="700"
    ohos:text_font="serif"
/>
```

组件常用的背景可以采用 XML 格式放置在 graphic 目录下。在 graphic 目录下创建 background_text.xml,在 XML 文件中定义文本的背景,代码如下:

```
<?xml version="1.0" encoding="utf-8"?>
<shape xmlns:ohos="http://schemas.huawei.com/res/ohos"
ohos:shape="rectangle">
    <corners ohos:radius="20"/>
    <solid ohos:color="#FF1992C3"/>
</shape>
```

最后,在 MainAbilitySlice.java 中,通过 setUIContent()加载该 XML 布局。代码如下:

```
public class MainAbilitySlice extends AbilitySlice {
    @Override
    public void onStart (Intent intent){
        Super.onStart(intent);
        Super.setUIContent(ResourceTable.Layout_text);
    }
}
```

运行效果如图 5-1 所示。

图 5-1　XML 创建 Text

(2) Java 代码中创建 Text。

打开 MainAbilitySlice.java 文件,找到 onStart()方法,创建 Text 组件,代码如下:

```
Text text = (Text) findComponentById
    (ResourceTable.Id_text);          //获取布局中的组件
text.setText("text 组件");             //设置文本
text.setTextSize(66);                 //设置文字大小
text.setTextColor(Color.WHITE);       //设置文字颜色
text.setWidth(300);
text.setHeight(150);
```

2. 设置 Text 组件样式

(1) 设置字体大小和颜色。

通常设置文本大小 text_size 和文本颜色 text_color 两个属性。text_size 为 float 类型，可以是浮点数值，其默认单位为 px；还可以是带 px/vp/fp 单位的浮点数值；还可以引用 float 资源。text_color 为 color 类型，可以直接设置色值，也可以引用 color 资源。代码如下：

```
ohos:text_size="66fp"
ohos:text_color="#FF0000"
```

(2) 设置字体风格和字重。

字体为 text_font 属性，可以设置的值包括 sans-serif、sans-serif-medium、HwChinese-medium、sans-serif-condensed、sans-serif-condensed-medium、monospace。italic 表示文本是否斜体字体，为 boolean 类型，可以直接设置 true/false，也可以引用 boolean 资源。text_weight 则表示字重。代码如下：

```
ohos:italic="true"
ohos:text_weight="700"
ohos:text_font="sans-serif"
```

(3) 设置文本对齐方式，换行和最大显示行数。

text_alignment 表示文本对齐方式，包含 left 文本靠左对齐，top 文本靠顶部对齐，right 文本靠右对齐，bottom 文本靠底部对齐，horizontal_center 文本水平居中对齐，vertical_center 文本垂直居中对齐，center 文本居中对齐，start 文本靠起始端对齐以及 end 文本靠结尾端对齐。multiple_lines 为多行模式设置，boolean 类型。max_text_lines 表示文本最大行数，为 integer 类型。代码如下：

```
ohos:text="Text 是一个文本组件"
ohos:text_alignment="horizontal_center"
ohos:multiple_lines="true"
ohos:max_text_lines="2"
```

运行效果如图 5-2 所示。

(4) 自动调节字体大小。

Text 对象可以实现根据文本长度来自动调整文本的字体大小和换行。通过设置 auto_font_size 来控制文本是否自动调整文本字体大小，该属性为 boolean 类型。效果如图 5-3

所示,代码如下:

图 5-2 文本效果

图 5-3 自动调节字体大小

```
ohos:multiple_lines="true"          //自动换行
ohos:max_text_lines="1"             //最大显示行数
ohos:auto_font_size="true"          //自动调节字体大小
```

(5) 跑马灯效果。

在文本过长的情况下,可以通过设置跑马灯效果来实现文本滚动显示。但在这里要注意的是,首先要关闭文本换行并且设置最大显示行数为1,一般默认即可。代码如下:

```
text.setTruncationMode(Text.TruncationMode.AUTO_SCROLLING);//自动滚动状态
text.setAutoScrollingCount(Text.AUTO_SCROLLING_FOREVER);
text.startAutoScrolling();                                  //启动跑马灯效果
```

5.1.2 图像组件

图像组件(Image)是用来在屏幕上显示图片的组件,各种图片的展示都需要该组件的使用。例如,在做网站首页时需要显示的背景图,动态加载显示用户头像以及朋友圈发布图片等场景,都可以使用 Image 图像组件进行显示。

1. 创建 Image

首先将所需图片资源 image.jpg 添加至 media 文件夹下,这里以"image.jpg"为例。在 image.xml 文件中声明组件,然后在 Java 文件中加载该 XML 布局。运行效果如图 5-4 所示,代码如下:

```
<Image
ohos:id="$+id:imageComponent"
ohos:height="match_parent"
ohos:width="match_parent"
ohos:image_src="$media:image"/>
```

或者用 Java 代码创建 Image 组件，在 MainAbilitySlice.java 文件中找到 onStart() 方法，创建 Image 组件，代码如下：

```
Image image = new Image(getContext());                    //创建 Image
image.setPixelMap(ResourceTable.Media_image);   //设置要显示的图片
DirectionalLayout layout = new DirectionalLayout(getContext());//创建布局
layout.addComponent(image); //Image 组件添加到布局中
super.setUIContent(layout);
```

2. 设置 Image 样式

（1）设置透明度。

通过设置 alpha 透明度属性来修改图片的透明度，效果如图 5-5 所示。代码如下：

```
ohos:alpha="0.5"        //设置透明度为 0.5
```

（2）设置缩放系数。

通过设置 X 轴和 Y 轴缩放系数来控制图片的缩放，效果如图 5-6 所示。代码如下：

图 5-4　创建 Image 组件　　　　图 5-5　透明度效果　　　　图 5-6　坐标轴缩放

```
ohos:scale_x="0.5"
ohos:scale_y="0.5"
```

（3）设置缩放方式。

当图片尺寸与 Image 尺寸不同时，可以根据不同的缩放方式对图片进行缩放。缩放方式 scale_mode 有下面几种：

- center 表示不缩放，按 Image 大小显示原图中间部分。
- zoom_center 表示原图按照比例缩放到与 Image 最窄边一致，并居中显示。
- zoom_start 表示原图按照比例缩放到与 Image 最窄边一致，并靠起始端显示。
- zoom_end 表示原图按照比例缩放到与 Image 最窄边一致，并靠结束端显示。
- stretch 表示将原图缩放到与 Image 大小一致。
- inside 表示将原图按比例缩放到与 Image 相同或更小的尺寸，并居中显示。
- clip_center 表示将原图按比例缩放到与 Image 相同或更大的尺寸，并居中显示。

图 5-7 缩放效果

例如，设置 Image 的宽为 300vp，高为 150vp，分别设置缩放方式为 zoom_start、zoom_center 和 zoom_end，为了突出效果，将 image 的背景色设为灰色，运行效果如图 5-7 所示。

代码如下：

```
<Image
    ohos:id="$+id:image"
    ohos:width="300vp"
    ohos:height="150vp"
    ohos:bottom_margin="20vp"
    ohos:image_src="$media:image"
    ohos:background_element="$graphic:background_image"
    ohos:scale_mode="zoom_start"/>
<Image
    ohos:id="$+id:image1"
    ohos:width="300vp"
    ohos:height="150vp"
    ohos:bottom_margin="20vp"
    ohos:image_src="$media:image"
    ohos:background_element="$graphic:background_image"
    ohos:scale_mode="zoom_center"/>
<Image
    ohos:id="$+id:image2"
    ohos:width="300vp"
    ohos:height="150vp"
    ohos:image_src="$media:image"
    ohos:background_element="$graphic:background_image"
    ohos:scale_mode="zoom_end"/>
```

（4）设置裁剪对齐方式。

当 Image 组件设定的尺寸小于图片实际尺寸时，可以对图片进行裁剪。通过设置 clip_alignment 样式来进行图片剪裁：left 表示左对齐裁剪；right 表示右对齐裁剪；top 表示顶部对齐裁剪；bottom 表示底部对齐裁剪；center 表示居中对齐裁剪。

例如，创建一个宽和高都为 200vp 的 Image 组件，设置左对齐 clip_alignment＝"left"，效果如图 5-8 所示。

图 5-8　裁剪效果

代码如下：

```
<Image
    ohos:id="$+id:image"
    ohos:width="200vp"
    ohos:height="200vp"
    ohos:layout_alignment="center"
    ohos:image_src="$media:image"
    ohos:clip_alignment="left"
/>
```

5.1.3　进度条组件

ProgressBar 用于显示内容或操作的进度。可以通过进度条查看一些功能操作的进度。使用场景：项目开发中通过设置数值改变进度条的样式，或者通过手动拖动改变进度条和进度值。

1. 创建 ProgressBar

在 layout 目录下的 XML 文件中创建一个 ProgressBar，同时在 Java 文件中加载该

XML 布局。运行效果如图 5-9 所示。代码如下：

```
<ProgressBar
    ohos:id="$+id:progressbar"
    ohos:orientation="horizontal"
    ohos:progress_width="40vp"
    ohos:height="120vp"
    ohos:width="match_parent"
    ohos:max="100"
    ohos:min="0"
    ohos:progress="60"
    ohos:progress_hint_text="60%"
    ohos:progress_hint_text_color="black"
/>
```

图 5-9　progressBar 效果

2. 设置 ProgressBar 样式

（1）设置 ProgressBar 的方向。

属性 orientation 表示排列方向，包含 horizontal 水平显示和 vertical 垂直显示。设置 ProgressBar 方向为水平，代码如下：

```
ohos:orientation="horizontal"
```

（2）设置 ProgressBar 的进度和提示文字。

设置当前进度 progress 为 integer 类型，可以直接设置整型数值，也可以引用 integer 资源。属性 progress_hint_text 表示进度提示文本，为 string 类型。progress_hint_text_color 表示进度提示文本颜色，为 color 类型。例如，设置进度为 60%，颜色为 black，代码如下：

```
ohos:progress="60"
ohos:progress_hint_text="60%"              //设置进度条数值
ohos:progress_hint_text_color="black"      //设置进度条数值颜色
```

(3) 设置最大值和最小值。

通过设置最大值 max 和最小值 min 来控制进度条，两个属性都为 integer 类型，可以直接设置整型数值，也可以引用 integer 资源。例如，设置最大值为 100，最小值为 0，在 XML 文件中设置：

```
ohos:max="100"
ohos:min="0"
```

或者在 Java 中设置：

```
progressBar.setMaxValue(100);
progressBar.setMinValue(0);
```

(4) 设置 ProgressBar 的颜色和分割线。

属性 background_instruct_element 表示背景色，progress_element 表示进度条颜色。两者都为 element 类型，可直接配置色值，也可引用 color 资源或引用 media/graphic 下的图片资源。代码如下：

```
ohos:progress_element="#FF9900"
ohos:background_instruct_element="#FF0000"
```

属性 divider_lines_enabled 表示分割线，为 boolean 类型，可以直接设置 true/false，也可以引用 boolean 资源。divider_lines_number 表示分割线数量，为 integer 类型，可以直接设置整型数值，也可以引用 integer 资源。运行效果如图 5-10 所示，如在 XML 文件中配置：

图 5-10 进度颜色和分割线效果

```
ohos:divider_lines_enabled="true"      //设置是否显示分割线
ohos:divider_lines_number="5"          //设置分割线个数
```

通过在 Java 文件中设置分割线颜色,代码如下:

```
progressBar.setDividerLineColor(Color.black);
```

5.1.4 圆形进度条

RoundProgressBar 继承自 ProgressBar,拥有 ProgressBar 的属性,在设置同样的属性时用法和 ProgressBar 一致,用于显示环形进度。

1. 创建 RoundProgressBar

在 layout 目录下的 XML 文件中创建一个 RoundProgressBar,同时在 Java 文件中加载该 XML 布局。运行效果如图 5-11 所示,代码如下:

```
<RoundProgressBar
    ohos:id="$+id:round_progress_bar"
    ohos:height="250vp"
    ohos:width="250vp"
    ohos:progress_width="15vp"
    ohos:progress="60"
    ohos:progress_color="blue"
    ohos:progress_hint_text="60%"
    ohos:progress_hint_text_color="#007DFF"
/>
```

2. 设置样式

组件 RoundProgressBar 的基础属性与组件 ProgressBar 保持一致,如设置进度条的大小、颜色、进度和提示文字等,都使用同样的属性进行控制,具体属性描述可以参考 5.1.3 节。这里主要讲解如何设置 RoundProgressBar 的开始和结束角度。

RoundProgressBar 的自有 XML 属性为 start_angle 和 max_angle。start_angle 表示圆形进度条的起始角度,max_angle 表示圆形进度条的最大角度。通过设置这两个属性值来控制进度条的开始和结束角度。例如,设置 start_angle 为 20,max_angle 为 340,运行效果如图 5-12 所示。代码如下:

```
ohos:start_angle="20"
ohos:max_angle="340"
```

5.1.5 时钟组件

时钟组件是 Text 的子类,所以可以使用 Text 的一些属性。

1. 创建 Clock

在 layout 目录下的 XML 文件中创建一个 Clock,同时在 Java 文件中加载该 XML 布

局。运行效果如图 5-13 所示。默认显示当前时间,而且时间是不断走动的。代码如下:

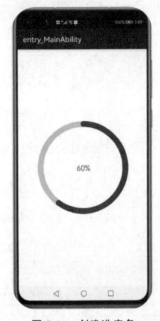
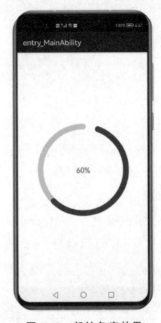

图 5-11　创建进度条　　　　图 5-12　起始角度效果　　　　图 5-13　创建 Clock

```
<Clock
    ohos:margin="40vp"
    ohos:height="match_content"
    ohos:width="match_content"
    ohos:text_size="40fp" />
```

2. 设置时间

(1) 设置 24 小时格式。

属性 time_zone 表示时区,包括 GMT(格林威治标准时间)、UTC(世界标准时间)、CST(美国,澳大利亚,古巴或中国的标准时间)、DST(夏令时)、PDT(太平洋夏季时间)。mode_24_hour 表示按照 24 小时显示的格式,值为指定的格式。示例效果如图 5-14 所示。代码如下:

```
<Clock
    ohos:id="$+id:clock"
    ohos:margin="40vp"
    ohos:height="500vp"
    ohos:width="250vp"
    ohos:time_zone="GMT"
    ohos:mode_24_hour="yyyy年MM月dd日 HH:mm:ss"
    ohos:multiple_lines="true"
    ohos:text_size="30fp" />
```

(2) 设置 12 小时格式。

在 Java 代码中设置时间的格式,通过 set24HourModeEnabled(boolean format24Hour) 方法设置时间是否按照 24 小时制进行显示,参数 false 表示不按 24 小时,true 表示按 24 小时,默认为 true。效果如图 5-15 所示,代码如下:

```
Clock clock = (Clock) findComponentById(ResourceTable.Id_clock);
//修改时钟组件展开方式,默认 24 小时。如要按 12 小时展示,先把 24 小时展示关闭
clock.set24HourModeEnabled(false);
clock.setFormatIn12HourMode("yyyy-MM-dd hh:mm:ss a");    //指定 12 小时的展示格式
```

3. 格式切换——24 小时制和 12 小时制之间的转换

通过单击按钮,将时钟组件中的显示方式在 24 小时制和 12 小时制之间切换。运行效果如图 5-16 所示。代码如下:

图 5-14　24 小时效果　　　　图 5-15　12 小时效果　　　　图 5-16　格式切换

```
<?xml version="1.0" encoding="utf-8"?>
<DirectionalLayout
    xmlns:ohos="http://schemas.huawei.com/res/ohos"
    ohos:height="match_parent"
    ohos:width="match_parent"
    ohos:orientation="vertical">
    <Clock
        ohos:id="$+id:clock"
        ohos:height="250vp"
        ohos:width="250vp"
        ohos:multiple_lines="true"
        ohos:text_size="35fp"
        ohos:text_alignment="center"
```

```xml
        ohos:mode_24_hour="yyyy-MM-dd  HH:mm:ss"
        ohos:layout_alignment="center" />
    <Button
        ohos:id="$+id:but"
        ohos:height="match_content"
        ohos:width="match_content"
        ohos:text="切换12小时制"
        ohos:text_size="30fp"
        ohos:text_color="white"
        ohos:background_element="#FF1582A4"
        ohos:top_margin="5vp"
        ohos:layout_alignment="center"/>
</DirectionalLayout>
```

Java 代码如下：

```java
public class MainAbilitySlice extends AbilitySlice implements
Component.ClickedListener {
    Clock clock;
    Button button;
    int flag = 0; //0表示24小时制,1表示12小时制
    @Override
    public void onStart(Intent intent) {
        super.onStart(intent);
        super.setUIContent(ResourceTable.Layout_clock);
        clock = (Clock) findComponentById(ResourceTable.Id_clock);
        button = (Button) findComponentById(ResourceTable.Id_but);
        button.setClickedListener(this); //给按钮添加一个单击事件
    }
    @Override
    public void onClick(Component component) {
        if(flag == 0){//改为12小时制
            clock.set24HourModeEnabled(false);//关闭24小时展示
            //12小时格式
            clock.setFormatIn12HourMode("yyyy-MM-dd hh:mm:ss a");
            button.setText("改为24小时制");
            flag = 1;
        }else if(flag == 1){ //改为24小时制
            clock.set24HourModeEnabled(true);//启用24小时计时制
            clock.setFormatIn24HourMode("yyyy-MM-dd  HH:mm:ss");
            button.setText("改为12小时制");
            flag = 0;
        }
    }
}
```

5.2 交互组件

5.2.1 按钮组件

Button 是一种常见的组件，单击可以触发对应的操作，通常由文本或图标组成，也可以由图标和文本共同组成。Button 无自有的 XML 属性，共有 XML 属性继承自 Text。

1. 创建 Button

（1）普通按钮。

普通按钮和其他按钮的区别在于不需要设置任何形状，只设置文本和背景颜色即可。在 layout 目录下的 XML 文件中创建 Button，同时在 Java 文件中加载该 XML 布局。设置按钮的背景形状、颜色，运行效果如图 5-17 所示。代码如下：

```
<Button
    ohos:id="$+id:button"
    ohos:width="200vp"
    ohos:height="100vp"
    ohos:text_size="30fp"
    ohos:text="普通按钮"
    ohos:text_color="white"
    ohos:background_element="$graphic:background_button"
    ohos:margin="25vp"
    ohos:padding="10vp" />
```

（2）带图标按钮。

在创建按钮的时候，可以通过 element_xxx 属性设置附带小图标。包含 element_left 文本左侧图标、element_top 文本上方图标、element_right 文本右侧图标、element_bottom 文本下方图标、element_start 文本开始方向图标、element_end 文本结束方向图标。

在"Project"窗口，打开"entry"→"src"→"main"→"resources"→"base"，右键单击"media"文件夹，粘贴进去一个 icon.jpg 图标。在 XML 文件中添加 Button 按钮，效果如图 5-18 所示，代码如下：

```
<Button
    ohos:id="$+id:button"
    ohos:width="match_content"
    ohos:height="match_content"
    ohos:text_size="30fp"
    ohos:text="带图标按钮"
    ohos:background_element="$graphic:background_button"
    ohos:margin="25vp"
    ohos:padding="10vp"
    ohos:element_padding="5vp"
    ohos:element_top="$media:icon"/>
```

(3) 其他形状的按钮。

按照按钮的形状,可以分为普通按钮、椭圆按钮、胶囊按钮、圆形按钮等。
- 普通按钮:和其他按钮的区别在于无须设置任何形状,只设置文本和背景颜色。
- 椭圆按钮:通过设置 background_element 来实现,background_element 的 shape 设置为椭圆(oval)。效果如图 5-19 所示,代码如下:

图 5-17　普通按钮　　　　图 5-18　带图标按钮　　　　图 5-19　按钮效果

```
<Button
    ohos:width="250vp"
    ohos:height="100vp"
    ohos:text_size="30fp"
    ohos:text="椭圆按钮"
    ohos:text_color="white"
    ohos:background_element="$graphic:oval_button"
    ohos:top_margin="30vp"
    ohos:left_margin="15vp"
    ohos:bottom_margin="15vp"
    ohos:right_padding="8vp"
    ohos:left_padding="8vp"/>
```

- 胶囊按钮:通过指定 background_element 的背景文件 capsule_button.xml 实现,在 capsule_button.xml 文件中设置 shape 为矩形(rectangle),并且指定 corners 元素的 ohos.radius 值。效果如图 5-20 所示,代码如下:

```
<Button
    ohos:width="250vp"
    ohos:height="100vp"
    ohos:text_size="30fp"
```

```
    ohos:text="胶囊按钮"
    ohos:text_color="white"
    ohos:background_element="$graphic:capsule_button"
    ohos:margin="25vp"
    ohos:padding="10vp"
/>
```

- 圆形按钮：和椭圆按钮的区别在于组件本身的宽度和高度需要相同。效果如图 5-21 所示，代码如下：

图 5-20　胶囊按钮　　　　　图 5-21　圆形按钮

```
<Button
    ohos:width="200vp"
    ohos:height="200vp"
    ohos:text_size="20fp"
    ohos:background_element="$graphic:oval_button"
    ohos:text="圆形按钮"
    ohos:text_color="white"
    ohos:margin="25vp"
    ohos:padding="10vp"
/>
```

2. 按钮单击事件

按钮的重要作用是当用户单击按钮时，会执行相应的操作或者界面出现相应的变化。实际上用户单击按钮时，Button 对象将收到一个单击事件。开发者可以自定义响应单击事件的方法。例如，通过创建一个 Component.ClickedListener 对象，然后调用 setClickedListener( ) 方法将其分配给按钮。代码如下：

```
Button button = (Button) findComponentById(ResourceTable.Id_button);
button.setClickedListener(new Component.ClickedListener() {
    @Override
    public void onClick(Component component) {
        //此处添加单击按钮后的事件处理逻辑
    }
});
```

5.2.2 切换按钮组件

ToggleButton 是一种用于显示开关状态的按钮控件，与 Switch 比较相似。ToggleButton 控件和 Button 控件的功能基本相同，ToggleButton 控件提供了可以表示"开/关"状态的功能。可以在连接/断开 Wi-Fi 或者打开/关闭手电筒等应用中使用。

1. 创建 ToggleButton

在 layout 目录下的 XML 文件中创建 ToggleButton，同时在 Java 文件中加载该 XML 布局。如创建一个普通开关，通过 width 和 height 设置控件宽度和高度，可设置为 match_content（按照内容自动伸缩）。运行效果如图 5-22 所示。代码如下：

```
<ToggleButton
    ohos:height="match_content"
    ohos:width="match_content"
    ohos:text_size="70fp"
    ohos:padding="10vp"
    ohos:background_element="$graphic:background"
/>
```

2. 设置样式

通过属性 text_color_on 设置选中时的文字颜色，text_color_off 设置未选中时的文字颜色，text_state_on 设置选中时文字，text_state_off 设置未选中时文字。例如，设置选中的文字为绿色"开"，未选中为蓝色"关"，未选中时的运行效果如图 5-23 所示。代码如下：

```
<ToggleButton
    ohos:height="match_content"
    ohos:width="match_content"
    ohos:text_state_off="关"
    ohos:text_state_on="开"
    ohos:text_color_off="blue"
    ohos:text_color_on="green"
    ohos:text_size="70fp"
    ohos:padding="10vp"
    ohos:background_element="$graphic:background"
/>
```

5.2.3 文本编辑组件

TextField 为一种文本输入框。TextField 的共有 XML 属性继承自 Text。TextField 的自有 XML 属性为 basement，即输入框基线，可直接配置色值，也可引用 color 资源或引用 media/graphic 下的图片资源。

1. 创建 TextField

在 layout 目录下的 XML 文件中创建一个 TextField，同时在 Java 文件中加载该 XML 布局。例如，创建一个宽为 300vp、高为 50vp 的文本框，效果如图 5-24 所示，具体组件属性描述代码如下：

图 5-22　普通开关

图 5-23　开关样式

图 5-24　文本框效果

```
<TextField
    ohos:id="$+id:textfield"
    ohos:height="50vp"
    ohos:width="300vp"
    ohos:text_size="15vp"
    ohos:margin="20vp"
    ohos:background_element=
    "$graphic:background_textfield"
    ohos:text_color="black"
    ohos:hint="请输入"
    ohos:multiple_lines="true"
    ohos:element_cursor_bubble=
    "$graphic:bubble"
    ohos:left_padding="10vp"
```

```
        ohos:right_padding="10vp"
        ohos:top_padding="8vp"
        ohos:bottom_padding="8vp"
        ohos:basement="#FF6E6A6A"
/>
```

2. 设置 TextField

(1) 设置 TextField 的背景。

首先在 layout 目录下 XML 文件中设置背景,代码如下:

```
ohos:background_element="$graphic:background_textfield"
```

graphic 目录下 XML 文件(如 background_textfield.xml)的代码示例如下:

```
<shape xmlns:ohos="http://schemas.huawei.com/res/ohos"
    ohos:shape="rectangle">
    <corners ohos:radius="40"/>
    <solid ohos:color="#72C6C2C2"/>
</shape>
```

(2) 设置 TextField 的提示文字。

通过属性 hint 设置提示文字,代码如下:

```
ohos:hint="请输入 "
```

(3) 设置 Bubble。

属性 element_cursor_bubble 表示文本的光标气泡图形,只有在可编辑的组件上可配置。为 element 类型,可直接配置色值,也可引用 color 资源或引用 media/graphic 下的图片资源。代码如下:

```
ohos:element_cursor_bubble="$graphic: bubble"
```

其中 bubble.xml 中代码如下:

```
<shape xmlns:ohos="http://schemas.huawei.com/res/ohos"
            ohos:shape="rectangle">
    <corners ohos:radius="40"/>
    <solid ohos:color="black"/>
    <stroke
        ohos:color="black"
        ohos:width="5"/>
</shape>
```

(4) 设置 TextField 的内边距。

文本框的内边距包含 left_padding 左边距、right_padding 右边距、top_padding 顶部边距和 bottom_padding 底部边距。如设置左右边距为 10vp,上下边距为 8vp,代码如下:

```
ohos:left_padding="10vp"
ohos:right_padding="10vp"
ohos:top_padding="8vp"
ohos:bottom_padding="8vp"
```

(5) 设置 TextField 的多行显示。

和 text 文本一样，通过属性 multiple_lines 来控制是否多行显示文本，为 boolean 类型。代码如下：

```
ohos:multiple_lines="true"
```

(6) 设置 TextField 不可用状态。

通过 TextField 的 setEnabled 属性来控制文本框是否可用，当设置成 false 后，文本框输入功能不可用。在 Java 代码中设置：

```
textField.setEnabled(false);
```

(7) 响应焦点变化。

通过 Java 代码设置文本框的响应焦点变化，代码如下：

```
textField.setFocusChangedListener((component, isFocused) -> {
    if(isFocused) { //获取焦点
} else { //失去焦点 }});
```

(8) 设置基线。

属性 basement 表示输入框基线，element 类型，可直接配置色值，也可引用 color 资源或引用 media/graphic 下的图片资源。如设置基线颜色为深灰色，代码如下：

```
ohos:basement="#FF6E6A6A"
```

5.2.4 单选组件

单选组件 RadioButton 是用于多选一操作的组件，它需要搭配 RadioContainer 使用，实现单选效果。RadioButton 的特点：①RadioButton 是圆形单选框；RadioContainer 是个可以容纳多个 RadioButton 的容器；②在 RadioContainer 中的 RadioButton 控件可以有多个，但同时有且仅有一个可以被选中；③每一个 RadioButton 有一个默认图标和文字组成。单选按钮允许用户在一组选项中选择一个选项。同一组中的单选按钮有互斥效果。

1. 创建 RadioButton

在 layout 目录下的 XML 文件中创建 RadioButton。如创建一个名为"单选按钮"的单选按钮，运行效果如图 5-25 所示，代码如下：

图 5-25 单选按钮

```xml
<RadioButton
    ohos:id="$+id:rb"
    ohos:height="match_content"
    ohos:width="match_content"
    ohos:text="单选按钮"
    ohos:text_size="50fp"
    ohos:text_color_on="blue"
    ohos:text_color_off="#808080"
/>
```

2. 样式设置

主要设置单选按钮选中与未选中两种不同状态下的字体颜色和标志样式。

（1）设置单选按钮的字体颜色。

通过设置选中时字体颜色属性 text_color_on 和未选中时字体颜色 text_color_off 来改变单选按钮字体颜色。代码如下：

```xml
ohos:text_color_on="#blue"
ohos:text_color_off="#808080"
```

（2）设置状态标志样式。

属性 check_element 表示状态标志样式，为 element 类型，可直接配置色值，也可引用 color 资源或引用 media/graphic 下的图片资源。效果如图 5-26 所示，代码如下：

```xml
ohos:check_element="$graphic:check"
```

在 graphic 文件中，创建 XML 文件，代码如下：

check.xml：
```xml
<state-container xmlns:ohos="http://schemas.huawei.com/res/ohos">
    <item
        ohos:element="$graphic:checked"
        ohos:state="component_state_checked"/>
    <item
        ohos:element="$graphic: empty"
        ohos:state="component_state_empty"/>
</state-container>
```
checked.xml：
```xml
<shape xmlns:ohos="http://schemas.huawei.com/res/ohos"
ohos:shape="rectangle">
    <solid ohos:color="blue"/>
    <corners ohos:radius="4"/></shape>
```
empty.xml：
```xml
<shape xmlns:ohos="http://schemas.huawei.com/res/ohos"
ohos:shape="rectangle">
    <solid ohos:color="gray"/>
    <corners ohos:radius="4"/> </shape>
```

图 5-26　状态标志样式

5.2.5　多选组件

多选组件 CheckBox 可以实现选中和取消选中的功能。CheckBox 和 Button 一样，也是一种按钮控件。它的优点在于，无须用户填写具体的信息，只需轻轻单击，即可选中程序希望用户填写的信息。如一个登录界面中，用它来判断需不需要记录用户密码。CheckBox 多选按钮之间不存在互斥关系，可以同时选择。

1. 创建 CheckBox

在 layout 目录下的 XML 文件中创建一个 CheckBox。效果如图 5-27 所示，代码如下：

图 5-27　复选框

```xml
<Checkbox
    ohos:id="$+id:checkbox"
    ohos:height="match_content"
    ohos:width="match_content"
    ohos:text="复选框"
    ohos:text_size="50vp"
    ohos:check_element="$graphic:check"
    ohos:text_color_on="#00AAEE"
    ohos:text_color_off="#000000"
/>
```

2. 设置 CheckBox

（1）在 XML 文件中配置 CheckBox 的选中和取消选中的状态标志样式。

和单选按钮一样，通过 check_element 来控制状态标志样式。代码如下：

```
ohos:check_element="$graphic: check"
```

（2）设置 CheckBox 的文字在选中和取消选中时的颜色。

通过设置选中时字体颜色属性 text_color_on 和未选中时字体颜色 text_color_off 来改变多选按钮字体颜色。代码如下：

```
ohos:text_color_on="#00AAEE"
ohos:text_color_off="#000000"
```

（3）设置 CheckBox 的选中状态。

在 Java 代码中，通过 setChecked()方法来控制 CheckBox 选中状态，代码如下：

```
checkbox.setChecked(true);
```

（4）设置不同状态之间的切换。

通过 CheckBox 的 toggle()方法来控制复选框不同状态之间的切换。如果当前为选中状态，那么将变为未选中；如果当前是未选中状态，将变为选中状态。代码如下：

```
checkbox.toggle();
```

（5）设置响应 CheckBox 状态变更的事件。

通过 CheckBox 的 setCheckedStateChangedListener()方法来控制 CheckBox 响应状态变更事件，代码如下：

```
//state 表示是否被选中
checkbox.setCheckedStateChangedListener((component, state) -> {
    //状态改变的逻辑
    ...
});
```

5.2.6 开关组件

Switch 是一种状态切换控件,它只有开和关两种状态,例如,在手机应用的设置界面做一些开关操作,或者某一个功能的开关操作。一般开关的文字是不显示的,只设置滑块和滑轨的颜色和样式。

1. 创建 Switch

在 layout 文件下创建布局资源文件 switch.xml,在 XML 文件中声明布局和组件,然后在 Java 文件中加载该 XML 布局。如创建一个高为 40vp,宽为 120vp 的开关,运行效果如图 5-28 所示,代码如下:

图 5-28　Switch 开关

```
<Switch
    ohos:id="$+id:switch_btn"
    ohos:height="40vp"
    ohos:width="120vp"
    ohos:text_size="20vp"
    ohos:text_state_off="OFF"
    ohos:text_state_on="ON"
/>
```

2. 设置样式

(1) 设置 Switch 状态。

在 Java 代码中通过设置 Switch 的 setChecked()方法来设置开关默认状态,代码如下:

```
Switch btnSwitch = (Switch) findComponentById
                (ResourceTable.Id_btn_switch);
//设置 Switch 默认状态
btnSwitch.setChecked(true);
```

(2) 设置 Switch 在开启和关闭时的文本。

属性 text_state_on 表示开启时显示的文本，text_state_off 表示关闭时显示的文本。例如，设置 Switch 在开启时显示"ON"，关闭时显示"OFF"，代码如下：

```
ohos:text_state_off="OFF"
ohos:text_state_on="ON"
```

(3) 设置响应 Switch 状态改变的事件。

在 Java 代码中设置响应 Switch 状态改变的事件，代码如下：

```
btnSwitch.setCheckedStateChangedListener(new AbsButton
        .CheckedStateChangedListener() {//回调处理 Switch 状态改变事件
    @Override
    public void onCheckedChanged(AbsButton button, boolean isChecked) {}
});
```

(4) 设置 Switch 的滑块和轨迹的样式。

属性 thumb_element 表示滑块样式，track_element 表示轨迹样式，两者都为 element 类型，可直接配置色值，也可引用 color 资源或引用 media/graphic 下的图片资源。示例效果如图 5-29 所示。

图 5-29　滑块、轨迹样式

代码如下:

```
ohos:thumb_element="$graphic:thumb"
ohos:track_element="$graphic:track"
```

graphic 目录下 XML 文件代码如下：

```
thumb.xml:
<state-container xmlns:ohos="http://schemas.huawei.com/res/ohos">
    <item ohos:state="component_state_checked"
    ohos:element="$graphic:thumb_element"/>
    <item ohos:state="component_state_empty"
    ohos:element="$graphic:thumb_element "/>
</state-container>
thumb_element.xml:
<shape xmlns:ohos="http://schemas.huawei.com/res/ohos" ohos:shape="oval">
    <solid ohos:color="#FFEDF5FC "/>
    <bounds
        ohos:top="0"
        ohos:left="0"
        ohos:right="40vp"
        ohos:bottom="40vp"/></shape>
track.xml:
<state-container xmlns:ohos="http://schemas.huawei.com/res/ohos">
    <item ohos:state="component_state_checked"
    ohos:element="$graphic:track_on"/>
    <item ohos:state="component_state_empty"
    ohos:element="$graphic:track_off"/>
</state-container>
track_on.xml:
<shape xmlns:ohos="http://schemas.huawei.com/res/ohos"
ohos:shape="rectangle">
    <solid ohos:color="#FF27DB27"/>
    <corners ohos:radius="25vp"/></shape>
track_off.xml:
<shape xmlns:ohos="http://schemas.huawei.com/res/ohos"
ohos:shape="rectangle">
    <solid ohos:color="#808080"/>
    <corners ohos:radius="25vp"/></shape>
```

5.3 高级组件

5.3.1 列表组件

列表组件（ListContainer）是用来呈现连续、多行数据的组件，包含一系列相同类型的列表项。

1. ListContainer 的使用方法

(1) ListContainer 使用步骤如下。

进行 XML 布局,创建控件 ListContainer;在对应的 AbilitySlice 中找到控件;设置数据源 List,Array,Map 等;创建 Provider,自定义类,继承 BaseItemProvider,重写 4 个必要的方法;设置 Provider,适配数据;添加监听 OnItemClickListener。

(2) 具体描述如下。

第一,在 layout 目录下,创建 listcontainer.xml 布局文件,在 XML 文件中创建 ListContainer。代码如下:

```
<ListContainer
    ohos:id="$+id:list_container"
    ohos:height="250vp"
    ohos:width="300vp"
    ohos:margin="35vp"
    ohos:layout_alignment="horizontal_center"
    ohos:background_element="#7EF5F0EA"
    ohos:shader_color="#95A8EAEC"/>
```

第二,在 layout 目录下新建每个 item 的布局文件 item.xml,作为 ListContainer 的子布局。代码如下:

```
<Text
    ohos:id="$+id:item_index"
    ohos:height="match_content"
    ohos:width="match_content"
    ohos:text="Item0"
    ohos:padding="15vp"
    ohos:text_size="20fp"
    ohos:layout_alignment="center"/>
```

第三,在 Java 代码中,创建 Item.java,作为 ListContainer 的数据包装类。代码如下:

```java
public class Item {
    private String name;
    public Item (String name) {this.name = name;}
    public String getName() { return name; }
    public void setName(String name) {this.name = name;}
}
```

第四,ListContainer 每一行可以为不同的数据,因此需要适配不同的数据结构,使其都能添加到 ListContainer 上。创建 ItemProvider.java,继承自 BaseItemProvider。必须重写的方法以及方法的作用如表 5-1 所示。

表 5-1 重写方法

方法	作用
int getCount()	返回填充的表项个数
Object getItem(int position)	根据 position 返回对应的数据
long getItemId(int position)	返回某一项的 id
Component getComponent(int position,Component covertComponent,ComponentContainer componentContainer)	根据 position 返回对应的界面组件

代码如下：

```java
public class ItemProvider extends BaseItemProvider {
    private List<Item> list;         //提供数据源
    private AbilitySlice slice;
    public ItemProvider(List<Item> list, AbilitySlice slice) {
        this.list = list;
        this.slice = slice; }
    @Override
    public int getCount() {return list == null ? 0: list.size();}
    @Override
    public Object getItem(int position) {
        if(list != null && position >= 0 && position < list.size()){
            return list.get(position);}
            return null;
    }
    @Override
    public long getItemId(int position) {return position;}
    @Override
    public Component getComponent(int position, Component convertComponent,ComponentContainer componentContainer) {
        final Component cpt;
        if(convertComponent == null) {
           cpt=LayoutScatter.getInstance(slice)
           .parse(ResouceTable.Layout_item_sample,null, false);
        } else { cpt = convertComponent; }
        Item item = list.get(position);
        Text text = (Text)cpt.findComponentById
        (ResourceTable.Id_item_index);
        text.setText(item.getName());
        return cpt;
    }
}
```

第五，在 MainAbilitySlice.java 文件中添加 ListContainer 的数据，并适配其数据结构。代码如下：

```java
public void onStart(Intent intent) {
    super.onStart(intent);
    super.setUIContent(ResourceTable.Layout_listcontainer);
    initListContainer();
}
private void initListContainer() {   //初始化
    ListContainer listContainer = (ListContainer)findComponentById
    (ResourceTable.Id_listcontainer);
    List<Item> list = getData();
    ItemProvider itemProvider = new ItemProvider(list, this);
    listContainer.setItemProvider(itemProvider);
}
private ArrayList<Item> getData() {   //初始化数据源
    ArrayList<Item> list = new ArrayList<>();
    for(int i = 0; i <= 8; i++) {list.add(new Item("Item" + i));}
    return list;
}
```

程序运行效果如图 5-30 所示。

2. ListContainer 的常用接口

（1）设置响应单击事件。

例如，单击 Item1，运行效果如图 5-31 所示，代码如下：

图 5-30　列表效果　　　　　　　　　图 5-31　单击事件

```
listContainer.setItemClickedListener ((container, component, position,
                                      id) -> {
    Item item = (Item) listContainer.getItemProvider().getItem(position);
    new ToastDialog(this).setText("你单击了" + item.getName())
        .setAlignment(LayoutAlignment.CENTER).show();});
```

(2) 设置响应长按事件。

效果如图 5-32 所示,代码如下:

```
listContainer.setItemLongClickedListener ((container, component, position,
                                          id) -> {
    Item item = (Item) listContainer.getItemProvider().getItem(position);
    new ToastDialog(this).setText("你长按了" + item.getName())
        .setAlignment(LayoutAlignment.CENTER).show();
    return false;});
```

3. ListContainer 的样式设置

(1) 设置 ListContainer 的布局方向。

属性 orientation 设置为"horizontal",表示横向布局;设置为"vertical",表示纵向布局。默认为纵向布局。代码如下:

可以在 XML 文件中静态设置:

```
ohos:orientation="horizontal"
```

也可以在 Java 代码中通过 setOrientation()方法动态设置布局方向:

```
listContainer.setOrientation(Component.HORIZONTAL);
```

(2) 设置 ListContainer 的开始和结束偏移量。

ListContainer 通过 setContentStartOffSet(int startOffset)、setContentEndOffSet(int endOffset)或者 setContentOffSet(int startOffset, int endOffset)来控制开始和结束的偏移量。例如,设置开始偏移量为 32,结束偏移量为 16,Java 代码如下:

```
listContainer.setContentOffSet(32, 16);
```

(3) 设置回弹效果。

属性 rebound_effect 表示开启/关闭回弹效果,为 boolean 类型,可以直接设置 true/false,也可以引用 boolean 资源。在 XML 文件中设置:

```
ohos:rebound_effect="true"
```

或在 Java 代码中通过 setReboundEffect(boolean enabled)方法设置是否启用回弹效果,代码如下:

```
listContainer.setReboundEffect(true);
```

在开启回弹效果后,可以调用 setReboundEffectParams() 方法调整回弹效果。

```
listContainer.setReboundEffectParams(40, 0.6f, 20);
```

(4) 设置着色器颜色。

属性 shader_color 表示着色器颜色,为 color 类型,可以直接设置色值,也可以引用 color 资源。运行效果如图 5-33 所示,代码如下:

```
ohos:shader_color="#95A8EAEC"
```

图 5-32 长按事件

图 5-33 着色器效果

5.3.2 标签列表组件

TabList 可以实现多个页签栏的切换,Tab 为某个页签。子页签通常放在内容区上方,展示不同的分类。页签名称应该简洁明了,清晰描述分类的内容。一般应用场景都是作为应用首页的二级菜单的多个分类。

1. 创建 TabList

在 layout 目录下的 XML 文件中创建一个 TabList,同时在 Java 文件中加载该 XML 布局。效果如图 5-34 所示,代码如下:

```
<TabList
    ohos:id="$+id:tablist"
    ohos:height="50vp"
    ohos:width="match_parent"
    ohos:normal_text_color="#000000"
```

```
ohos:orientation="horizontal"
ohos:selected_tab_indicator_color="red"
ohos:selected_tab_indicator_height="2vp"
ohos:selected_text_color="red"
ohos:tab_length="60vp"
ohos:tab_margin="20vp"
ohos:text_alignment="center"
ohos:text_size="20fp"/>
```

2. 设置样式

（1）设置默认状态和选中状态的文本颜色和页签的颜色。

属性 normal_text_color 表示未选中的文本颜色，selected_text_color 表示选中的文本颜色，selected_tab_indicator_color 表示选中页签的颜色，三者都为 color 类型，可以直接设置色值，也可以引用 color 资源。selected_tab_indicator_height 表示选中页签的高度，为 float 类型。代码如下：

```
ohos:normal_text_color="#000000"
ohos:selected_text_color="#red"
ohos:selected_tab_indicator_color="#red"
ohos:selected_tab_indicator_height="2vp"
```

（2）TabList 中添加 Tab。

在 Java 代码中，获取布局中的 TabList，添加两个 Tab，分别为推荐和视频，代码如下：

```
TabList tabList = (TabList)findComponentById(ResourceTable.Id_tablist);
TabList.Tab tab = tabList.new Tab(getContext());
tab.setText("推荐");
tabList.addTab(tab);
TabList.Tab tab2 = tabList.new Tab(getContext());
tab2.setText("视频");
tabList.addTab(tab2);
```

（3）设置 Tab 的布局。

通过 tab_margin 和 tab_length 来控制 Tab 的布局。属性 tab_margin 表示页签间距，tab_length 表示页签长度，都为 float 类型。例如，设置页签间距为 20vp，长度为 60vp，在 XML 文件中设置 Tab 的布局，代码如下：

```
ohos:tab_margin="20vp"
ohos:tab_length="60vp"
```

（4）设置 fixedMode。

fixedMode 表示固定模式，默认为 false，该模式下 TabList 的总宽度是各 Tab 宽度的总和，若固定了 TabList 的宽度，当超出可视区域，则可以通过滑动 TabList 来显示。如果设置为 true，TabList 的总宽度将与可视区域相同，各个 Tab 的宽度也会根据 TabList 的宽度

而平均分配,该模式适用于 Tab 较少的情况。代码如下:

```
TabList.setFixedMode(true);
```

(5) 新增 Tab。

在 Java 代码中通过 addTab()方法在指定位置添加 Tab,例如,在"推荐"和"视频"之间的页签中新增"音乐"页签,运行效果如图 5-35 所示,代码如下:

```
TabList.Tab tab3 = tabList.new Tab(getContext());
tab3.setText("音乐");
tabList.addTab(tab3, 1, true);//1 表示位置, true 表示选中
```

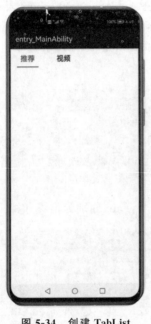
图 5-34　创建 TabList

图 5-35　新增 Tab

(6) 响应焦点变化。

在 Java 代码中设置 TabList 响应焦点变化,代码如下:

```
tabList.addTabSelectedListener(new TabList.TabSelectedListener() {
    @Override
    public void onSelected(TabList.Tab tab) {//Tab 从未选中变为选中时的回调}
    @Override
    public void onUnselected(TabList.Tab tab) {//Tab 从选中变为未选中时的回调}
    @Override
    public void onReselected(TabList.Tab tab) {//Tab 被选中,再次单击时的回调}
});
```

(7) TabList 常用接口。

TabList 常用接口如表 5-2 所示。

表 5-2 常用接口

接口名	作用	接口名	作用
getSelectedTab	返回选中的 Tab	getTabAt	获取某个 Tab
getSelectedTabIndex	返回选中的 Tab 的位置索引	removeTab	移除某个位置的 Tab
getTabCount	获取 Tab 的个数	setOrientation	设置横或竖方向

5.3.3 滑动选择器组件

Picker 提供了滑动选择器,允许用户从预定义范围中进行选择。一般都是在三个以上选项的时候才建议使用该功能。

1. 使用 Picker

(1) 创建 Picker。

在 layout 目录下的 XML 文件中创建一个 Picker,同时在 Java 文件中加载该 XML 布局。效果如图 5-36 所示,代码如下:

```
<Picker
    ohos:id="$+id:test_picker"
    ohos:height="match_content"
    ohos:width="300vp"
    ohos:layout_alignment="horizontal_center"
    ohos:normal_text_size="16fp"
    ohos:normal_text_color="gray"
    ohos:selected_text_size="16fp"
    ohos:selected_text_color="black"
    ohos:bottom_line_element="black"
    ohos:top_line_element="black"
    ohos:wheel_mode_enabled="true"
    ohos:selected_normal_text_margin_ratio="5.0"
    ohos:shader_color="#461E90FF"/>
```

(2) 设置 Picker 的取值范围。

在 Java 代码中通过 setMinValue()和 setMaxValue()方法控制 Picker 的取值范围,例如,设置取值范围为 0~8,代码如下:

```
Picker picker = (Picker) findComponentById(ResourceTable.Id_test_picker);
picker.setMinValue(0);         //设置选择器中的最小值
picker.setMaxValue(8);         //设置选择器中的最大值
```

(3) 响应选择器变化。

在 Java 代码中通过 setValueChangedListener()方法来响应选择器变化,代码如下:

```
picker.setValueChangedListener((picker1, oldVal, newVal) -> {
    //oldVal 为上一次选择的值;newVal 为最新选择的值
});
```

(4) 格式化 Picker 的显示。

通过 Picker 的 setFormatter(Formatter formatter)方法,用户可以将 Picker 选项中显示的字符串修改为特定的格式。运行效果如图 5-37 所示,代码如下:

```
picker.setFormatter(i -> {
    String value;
    switch (i) {
        case 0:value = "音乐";break;
        case 1:value = "游泳";break;
        case 2:value = "唱歌";break;
        case 3:value = "爬山";break;
        default:value = "" + i;
    }
    return value;
});
```

图 5-36　创建 Picker

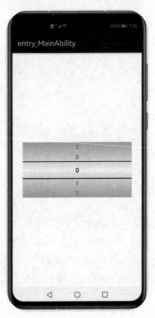

图 5-37　格式化效果

(5) 设置要显示的字符串数组。

对于不直接显示数字的组件,可以通过 setDisplayedData()方法设置字符串与数字一一对应。字符串数组长度必须等于取值范围内的值总数。同时,需要注意的一点是,该方法在使用过程中会覆盖 setFormatter()方法。Java 代码如下:

```
picker.setDisplayedData(new String[]{"音乐","游泳","唱歌","爬山"});
```

2. 样式设置

(1) 文本相关属性。

属性 normal_text_color 表示未选中的文本颜色,selected_text_color 表示选中的文本

颜色，都为 color 类型，可直接设置色值，也可以引用 color 资源。normal_text_size 表示未选中的文本大小，selected_text_size 表示选中的文本大小，都为 float 类型，可以是浮点数值，其默认单位为 px；也可以是带 px/vp/fp 单位的浮点数值；也可以引用 float 资源。代码如下：

```
ohos:normal_text_size="16fp"
ohos:normal_text_color="gray"
ohos:selected_text_size="16fp"
ohos:selected_text_color="black"
```

(2) 设置所选文本的上下边框。

属性 top_line_element 表示选中项的上边框，bottom_line_element 表示选中项的下边框，都为 element 类型，可直接配置色值，也可以引用 color 资源或引用 media/graphic 下的图片资源。代码如下：

```
ohos:bottom_line_element="black"
ohos:top_line_element="black"
```

(3) 设置 Picker 的着色器颜色。

属性 shader_color 表示着色器颜色，为 color 类型，可直接设置色值，也可以引用 color 资源。在 XML 文件中设置：

```
ohos:shader_color="#461E90FF"
```

或者在 Java 代码中通过 setShaderColor()方法设置着色器颜色，代码如下：

```
picker.setShaderColor(new Color(Color.getIntColor("#461E90FF")));
```

(4) 设置 Picker 中已选文本边距与常规文本边距的比例。

属性 selected_normal_text_margin_ratio 已选文本边距与常规文本边距的比例，为 float 类型，可直接设置浮点数值，也可以引用 float 浮点数资源。取值需>0.0f，默认值为 1.0f。例如，设置比例为 5.0，在 XML 文件中代码如下：

```
ohos:selected_normal_text_margin_ratio="5.0"
```

或者在 Java 代码中设置：

```
picker.setSelectedNormalTextMarginRatio(5.0f);
```

(5) 设置选择轮模式。

该模式决定 Picker 是否循环显示数据。属性 wheel_mode_enabled 表示选择轮是否循环显示数据，为 boolean 类型，可直接设置 true/false，也可以引用 boolean 资源。在 XML 文件中代码如下：

```
ohos:wheel_mode_enabled="true"
```

或者在 Java 代码中设置：

```
picker.setWheelModeEnabled(true);
```

5.3.4 日期选择器组件

DatePicker 组件的功能主要供用户选择日期。

1. 使用 DatePicker

（1）在 XML 文件中创建 DatePicker。

在 layout 目录下的 XML 文件中创建一个 DatePicker，同时在 Java 文件中加载该 XML 布局。运行效果如图 5-38 所示，代码如下：

图 5-38　DatePicker

```
<DatePicker
    ohos:id="$+id:date_pick"
    ohos:height="match_content"
    ohos:width="300vp"
    ohos:normal_text_color="gray"
    ohos:normal_text_size="20fp"
    ohos:selected_text_color="black"
    ohos:selected_text_size="20fp"
    ohos:operated_text_color="blue"
    ohos:selected_normal_text_margin_ratio="10"
    ohos:wheel_mode_enabled="true"
    ohos:top_line_element="black"
```

```
    ohos:bottom_line_element="black"
    ohos:shader_color="#7C69BEB6"
/>
```

(2) 获取当前选择日期。

在 Java 代码中通过 getDayOfMonth()、getMonth() 和 getYear() 方法获取日/月/年，DatePicker 默认选择当前日期。代码如下：

```
DatePicker datePicker = (DatePicker) findComponentById
                        (ResourceTable.Id_date_pick);
int day = datePicker.getDayOfMonth();
int month = datePicker.getMonth();
int year = datePicker.getYear();
```

(3) 响应日期改变事件。

在 Java 代码中通过 setValueChangedListener() 来控制响应日期改变事件，代码如下：

```
Text selectedDate = (Text)findComponentById(ResourceTable.Id_date);
datePicker.setValueChangedListener(new DatePicker.ValueChangedListener() {
    @Override
    public void onValueChanged(DatePicker datePicker, int year,
    int monthOfYear, int dayOfMonth) {
    selectedDate.setText(String.format("时间: %02d/%02d/%02d",
    year, monthOfYear, dayOfMonth);
    }
});
```

在 XML 文件中添加 Text 显示选择日期：

```
<Text
    ohos:id="$+id:date"
    ohos:height="match_content"
    ohos:width="match_parent"
    ohos:hint="date"
    ohos:padding="4vp"
    ohos:text_size="20fp"
    ohos:left_margin="20vp">
</Text>
```

(4) 设置日期的范围。

如果需要对 DatePicker 的日期选择范围进行设置，可以通过设置属性 min_date 和 max_date 来实现。设置的值为日期对应的 Unix 时间戳。

属性 min_date 表示最小日期，为 long 类型，可以直接设置长整型值，也可以引用 string 资源。如在 XML 文件中设置：

```
ohos:min_date="1627747200"
```

属性 max_date 表示最大日期,为 long 类型,可以直接设置长整型值,也可以引用 string 资源。如在 XML 文件中设置:

```
ohos:max_date="1630339200"
```

(5) 固定年/月/日

属性 year_fixed 表示年份是否固定,month_fixed 表示月份是否固定,day_fixed 表示日期是否固定,都为 boolean 类型,可以直接设置 true/false,也可以引用 boolean 资源。例如,设置时间中年/月/日固定,在 XML 文件中代码如下:

```
ohos:year_fixed="true"
ohos:month_fixed="true"
ohos:day_fixed="true"
```

2. 样式设置

(1) 文本相关属性。

属性 normal_text_size 表示未选中文本的大小,selected_text_size 表示选中文本的大小,均为 float 类型。normal_text_color 表示未选中文本的颜色,selected_text_color 表示选中文本的颜色,均为 color 类型,可以直接设置色值,也可以引用 color 资源。代码如下:

```
ohos:normal_text_color="gray"
ohos:normal_text_size="20fp"
ohos:selected_text_color="black"
ohos:selected_text_size="20fp"
```

属性 operated_text_color 表示操作项的文本颜色,为 color 类型,可以直接设置色值,也可以引用 color 资源。在 XML 文件中设置:

```
ohos:operated_text_color="blue"
```

(2) 设置 DatePicker 中已选文本边距与常规文本边距的比例。

属性 selected_normal_text_margin_ratio 表示已选文本边距与常规文本边距的比例,为 float 类型,可以直接设置浮点数值,也可以引用 float 资源。取值需>0.0f,默认值为 1.0f。例如,设置比例为 10,在 XML 文件中代码如下:

```
ohos:selected_normal_text_margin_ratio="10"
```

(3) 设置选择轮模式。

属性 wheel_mode_enabled 表示选择轮是否循环显示数据,为 boolean 类型,可以直接设置 true/false,也可以引用 boolean 资源。代码如下:

```
ohos:wheel_mode_enabled="true">
```

(4) 设置选中日期的上下边框。

属性 top_line_element 表示选中项的上边框,bottom_line_element 表示选中项的下边

框，都为 element 类型，可直接配置色值，也可引用 color 资源或引用 media/graphic 下的图片资源。在 XML 文件中设置：

```
ohos:top_line_element="black"
ohos:bottom_line_element="black"
```

（5）设置着色器颜色。

属性 shader_color 表示着色器颜色，为 color 类型，可以直接设置色值，也可以引用 color 资源。在 XML 文件中代码如下：

```
ohos:shader_color="#7C69BEB6"
```

5.3.5 时间选择器组件

TimePicker 组件的功能主要供用户选择时间。

1. 使用 TimePicker

（1）创建 TimePicker。

在 layout 目录下的 XML 文件中创建一个 TimePicker，同时在 Java 文件中加载该 XML 布局。运行效果如图 5-39 所示，代码如下：

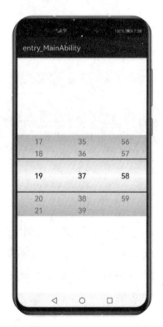

图 5-39　TimePicker

```
<TimePicker
    ohos:id="$+id:time_picker"
    ohos:height="match_content"
    ohos:width="match_parent"
    ohos:normal_text_color="gray"
```

```
    ohos:normal_text_size="20fp"
    ohos:selected_text_color="black"
    ohos:selected_text_size="20fp"
    ohos:operated_text_color="blue"
    ohos:selected_normal_text_margin_ratio="10"
    ohos:shader_color="#6A54AEAE"
    ohos:bottom_line_element="black"
    ohos:top_line_element="black"
    ohos:mode_24_hour="false"
    ohos:am_pm_order="right"
/>
```

(2) 获取时间。

在 Java 代码中通过 getHour()、getMinute() 和 getSecond() 方法获取时/分/秒。代码如下:

```
TimePicker timePicker = (TimePicker) findComponentById
                       (ResourceTable.Id_time_picker);
int hour = timePicker.getHour();
int minute = timePicker.getMinute();
int second = timePicker.getSecond();
```

(3) 响应时间改变事件。

在 Java 代码中通过 setTimeChangedListener() 来控制响应时间改变事件,代码如下:

```
timePicker.setTimeChangedListener(new TimePicker.TimeChangedListener() {
    @Override
    public void onTimeChanged(TimePicker timePicker, int hour, int minute,
    int second) {}
});
```

2. 显示样式配置

(1) 设置字体属性。

属性 normal_text_size 表示未选中文本的大小,selected_text_size 表示选中文本的大小,都为 float 类型。normal_text_color 表示未选中文本的颜色,selected_text_color 表示选中文本的颜色,operated_text_color 表示操作项的文本颜色,都为 color 类型,可以直接设置色值,也可以引用 color 资源。代码如下:

```
ohos:normal_text_color="gray"
ohos:normal_text_size="20fp"
ohos:selected_text_color="black"
ohos:selected_text_size="20fp"
ohos:operated_text_color="blue"
```

(2) 设置 TimePicker 中已选文本边距与常规文本边距的比例。

属性 selected_normal_text_margin_ratio 表示已选文本边距与常规文本边距的比例,为

float 类型,可以直接设置浮点数值,也可以引用 float 资源。取值需大于 0.0f,默认值为 1.0f。

如设置比例为 10,代码如下:

```
ohos:selected_normal_text_margin_ratio="10"
```

(3) 设置着色器颜色。

属性 shader_color 表示着色器颜色,为 color 类型,可以直接设置色值,也可以引用 color 资源。代码如下:

```
ohos:shader_color="#6A54AEAE"
```

(4) 设置选中时间的上下边框。

属性 top_line_element 表示选中项的上边框,bottom_line_element 表示选中项的下边框,都为 element 类型,可直接配置色值,也可引用 color 资源或引用 media/graphic 下的图片资源。如设置下边框颜色,代码如下:

```
ohos:bottom_line_element="black"
ohos:top_line_element="black"
```

(5) 设置 12 小时制下显示样式。

属性 am_pm_order 表示在 12 小时制显示的情况下,控制上午下午排列顺序,取值包含 start 表示 am/pm 列的位置靠时间选择器起始端,end 表示 am/pm 列的位置靠时间选择器结束端,left 表示 am/pm 列的位置靠时间选择器左侧,right 表示 am/pm 列的位置靠时间选择器右侧。mode_24_hour 表示是否 24 小时制显示,为 boolean 类型,可以直接设置 true/false,也可以引用 boolean 资源。

AM/PM 默认置于左侧,如设置位于右边,运行效果如图 5-40 所示,代码如下:

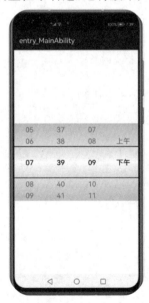

图 5-40 12 小时显示

```
ohos:mode_24_hour="false"
ohos:am_pm_order="right"
```

3. 范围选择设置

(1) 设置隐藏或显示时分秒。

在 Java 代码中通过 showHour()、showMinute()、showSecond()方法来控制是否隐藏时分秒,如隐藏时分秒,代码如下:

```
timePicker.showHour(false);
timePicker.showMinute(false);
timePicker.showSecond(false);
```

(2) 设置 TimePicker 的 selector 是否可以滑动。

在 Java 代码中通过 enableHour()、enableMinute()、enableSecond()方法来控制 selector 是否可以滑动,如设置时分秒 selector 无法滑动选择,代码如下:

```
timePicker.enableHour(false);
timePicker.enableMinute(false);
timePicker.enableSecond(false);
```

5.3.6 滚动视图组件

ScrollView 是一种带滚动功能的组件,它通过采用滚动的方式在有限的区域内显示更多的内容。

1. 创建 ScrollView

在 layout 目录下的 XML 文件中创建 ScrollView,ScrollView 的展示需要布局支持,此处以 DirectionalLayout 为例,运行效果如图 5-41 所示,代码如下:

```
<ScrollView
    ohos:id="$+id:scrollview"
    ohos:height="200vp"
    ohos:width="200vp"
    ohos:background_element="#FFDEAD"
    ohos:top_margin="32vp"
    ohos:bottom_padding="16vp"
    ohos:layout_alignment="horizontal_center">
    <DirectionalLayout
        ohos:height="match_content"
        ohos:width="match_content"
        ohos:orientation="horizontal">
        <Image
            ohos:width="300vp"
            ohos:height="match_content"
            ohos:top_margin="16vp"
```

```
            ohos:image_src="$media:image"
        />
    </DirectionalLayout>
</ScrollView>
```

2. 设置 ScrollView

(1) 根据像素数平滑滚动。

添加按钮,单击后根据像素数平滑滚动 ScrollView。

```
<Button
    ohos:id="$+id:btn"
    ohos:height="match_content"
    ohos:width="match_content"
    ohos:background_element="blue"
    ohos:layout_alignment="center"
    ohos:padding="10vp"
    ohos:text="Scroll By Y:300"
    ohos:text_color="white"
    ohos:text_size="20fp"
    ohos:top_margin="16vp"
/>
```

在 Java 代码中设置按钮单击事件并开启滚动,运行效果如图 5-42 所示,代码如下:

图 5-41　ScrollView 效果

图 5-42　滚动效果

```
ScrollView scrollView = (ScrollView) findComponentById
                (ResourceTable.Id_scrollview);
```

```
Button button = (Button)findComponentById(ResourceTable.Id_btn);
button.setClickedListener(component -> {
    scrollView.fluentScrollByY(300);
});
```

(2) 平滑滚动到指定位置。

在 Java 代码中通过 fluentScrollTo(int x，int y)方法，fluentScrollXTo(int x)方法或 fluentScrollYTo(int y)方法设置沿坐标轴将内容平滑地滚动指定数量的像素（单位：px）。例如，根据 Y 轴平滑滚动到 500px 位置处，代码如下：

```
scrollView.fluentScrollYTo(500);
```

(3) 设置布局方向。

ScrollView 自身没有设置布局方向的属性，所以需要在其子布局中设置。以横向布局 horizontal 为例，代码如下：

```
<ScrollView >
    <DirectionalLayout
        ohos:orientation="horizontal">
    </DirectionalLayout>
</Scrollview>
```

(4) 设置回弹效果。

属性 rebound_effect 表示回弹效果，为 boolean 类型，可以直接设置 true/false，也可以引用 boolean 资源。例如，在 XML 文件中设置：

```
ohos:rebound_effect="true"
```

或者在 Java 代码中设置：

```
scrollView.setReboundEffect(true);
```

(5) 设置拉伸匹配效果。

属性 match_viewport 表示是否拉伸匹配，为 boolean 类型，可以直接设置 true/false，也可以引用 boolean 资源。该属性在 ScrollView 的子组件无法填充满 ScrollView 时使用，默认为 false，子组件按照自身设置大小展示，设置为 true 时，子组件填充满 ScrollView。例如，在 XML 文件中设置：

```
ohos:match_viewport="true"
```

或者在 Java 代码中设置：

```
scrollView.setMatchViewportEnabled(true);
```

第 6 章 对 话 框

本章介绍如何在 HarmonyOS 中弹出对话框。对话框本质上就是窗口，开发者只需要调用 HarmonyOS 提供的现成的对话框类（CommonDialog 类），就可以实现各种样式的对话框。

通过阅读本章，读者可以掌握：
- 显示对话框。
- 为对话框添加一个或多个按钮。
- 自动关闭对话框。
- 定制对话框。
- 显示 Toast 信息框。
- Popup 对话框的使用。

6.1 普通对话框

使用 CommonDialog 类可以实现一个对话框，该类有很多 API，可以定制不同风格的对话框，本节将详细介绍 CommonDialog 类的用法。

6.1.1 显示一个简单的对话框

标题（title）和内容（content）是对话框包含的最基本的两个信息，可以通过调用 CommonDialog 类的 setTitleText()方法和 setContentText()方法来设置对话框的标题和内容。Show()方法可以显示对话框，具体代码如下：

```
CommonDialog cd =new CommonDialog(this);
cd.setTitleText("消息");
cd.setContentText("这是普通对话框");
cd.show();
```

运行程序，就会弹出如图 6-1 所示的对话框。

CommonDialog 是一种在弹出框消失之前，用户无法操作其他界面内容的对话框。这样的对话框在实际应用中存在很多限制，由于该对话框没有"关闭""确定"这样的按钮，所以关闭对话框的唯一方法是按回退键。在下一节将介绍如何为对话框添加"关闭"按钮。

图 6-1　普通对话框

6.1.2　为对话框添加"关闭"按钮

可以通过调用 CommonDialog 类的 setButton()方法为对话框添加一个或多个按钮,通过设置按钮区的按钮样式,可自定义按钮的位置、文本及相关单击事件。该方法的接口如下：

```
setButton(int buttonNum, String text, IDialog.ClickedListener listener);
```

setButton()方法的参数描述如下。

buttonNum：按钮的索引。在多个按钮共用同一个单击事件的情况下,可以通过该参数来区分哪一个按钮进行了单击事件。

text：在按钮上显示的文本。

listener：单击按钮响应的事件监听器。

需要用单击事件监听器来响应单击按钮的动作,否则新添加的按钮不会响应任何动作。例如,可以调用 CommonDialog 类的 destroy()方法来实现对话框的关闭。代码如下：

```
CommonDialog cd = new CommonDialog(this);
cd.setTitleText("消息");
cd.setContentText("这是带关闭按钮的对话框");
cd.setButton(0, "关闭", newIDialog.ClickedListener(){
    @Override
    Public void onClick(IDialog iDialog, int i){
```

```
        cd.destroy();
    }
});
cd.show();
```

运行程序,会弹出如图 6-2 所示的对话框,单击"关闭"按钮后就会关闭对话框。

图 6-2 带关闭按钮的对话框

6.1.3 为对话框添加多个按钮

使用 setButton()方法理论上可以为对话框添加多个按钮,但目前最多只能为对话框添加两个按钮,超过两个按钮,多余的按钮将被忽略。下面的代码为对话框添加了"确定"按钮和"取消"按钮。

```
CommonDialog cd = new CommonDialog(this);
cd.setTitleText("消息");
cd.setContentText("这是带两个按钮的对话框");
IDialog.ClickedListener cl = new IDialog.ClickedListener(){
    @Override
    Public void onClick(IDialog iDialog, int i) {
        cd.destroy();
    };
    cd.setButton(0, "确定", cl);
    cd.setButton(1, "取消", cl);
    cd.show();
}
```

运行程序,会弹出如图 6-3 所示的对话框。

图 6-3 带两个按钮的对话框

本例中"确定"按钮和"取消"按钮共用了一个单击事件监听器。如果对话框指定两个按钮,setButton()方法所设置的索引必须是连续的,例如,不能使用下面的代码对话框指定两个按钮。

```
cd.setButton(1, "确定", cl);
cd.setButton(3, "取消", cl);
```

在这段代码中,按索引分别为 1 和 3,所以索引为 3 的按钮会被忽略。

6.1.4 调整按钮的尺寸

前面弹出的按钮在水平方向上是充满整个屏幕的,使用 setSize()方法可以重新设置对话框的尺寸(宽度和高度)。该方法的原型如下:

```
public BaseDialog setSize(int width, int height);
```

由于移动设备和屏幕千差万别,因此在设计 App 时通常不会采用物理像素,而是采用虚拟像素,例如,HarmonyOS 的虚拟像素单位是 vp。不管设备实际的分辨率是多少,使用虚拟像素的设备的分辨率都是一样的。当 App 运行在实际的设备上时,系统会自动根据屏幕密度将虚拟像素转换为物理像素。如果屏幕密度是 200dpi,那么 1vp=1px,这里 px 表示物理像素。所以如果 App 在拥有 200dpi 屏幕密度的设备上运行,那么 100vp 就相当于 100px 了。如果屏幕密度是 400dpi,那么 200vp 就会转换为 400px。这样可以保证组件的相对位置和相对尺寸不变。

完整代码如下：

```
CommonDialog cd = new CommonDialog(this);
cd.setTitleText("消息");
cd.setContentText("调整尺寸的对话框");
int density = getResourceManager().getDeviceCapability().screenDensity
            / DeviceCapability.SCREEN_MDPI;
int width = 300 * density;
int height= 100 * density;
cd.setSize(width, height);
IDialog.ClickedListener cl=new IDialog.ClickedListener() {
    @Override
    public void onClick(IDialog iDialog, int i) {
        cd.destroy();
    }
};
cd.setButton(0, "确定", cl);
cd.setButton(1, "取消", cl);
cd.show();
```

运行程序，会显示如图 6-4 所示的对话框。

图 6-4　调整尺寸的对话框

6.1.5 自动关闭对话框

除了之前通过给按钮添加 destory() 的监听事件,单击按钮关闭对话框,CommonDialog 类还允许单击非对话框区域自动关闭对话框,这里需要使用 setAutoClosable() 方法允许自动关闭对话框。完整代码如下:

```
CommonDialog cd = new CommonDialog(this);
cd.setTitleText("消息");
int density = getResourceManager().getDeviceCapability().screenDensity /
              DeviceCapability.SCREEN_MDPI;
int width = 300 * density;
int height = 100 * density;
cd.setContentText("自动关闭对话框");
cd.setAutoClosable(true);
cd.setButton(0, "关闭", null);
cd.show();
```

运行程序,会显示如图 6-5 所示的对话框。此时单击关闭按钮不能将对话框关闭,因为未给关闭按钮添加监听事件。当单击非对话框区域时,对话框将自动关闭。

图 6-5　自动关闭对话框

6.2 定制对话框

如果对默认的对话框样式不太满意,setContentCustomComponent()方法可以定制对话框,该方法可以指定一个视图(通常直接使用布局文件装载),可以用该视图完全替换标题、内容和按钮。下面的代码实现了一个自定义样式的对话框,在布局文件中自定义了对话框标题的颜色和对话框中内容的颜色。

```
Component dl = LayoutScatter.getInstance(getContext()).
        parse(ResourceTable.Layout_custom_dial og, null, false);
Text text = (Text) dl.findComponentById(ResourceTable.Id_content);
//设置对话框内容
text.setText("定制对话框");
CommonDialog cd = new CommonDialog(this);
cd.setAlignment(LayoutAlignment.CENTER);
int density = getResourceManager().getDeviceCapability().screenDensity
        / DeviceCapability.SCREEN_MDPI;
int width = 600 * density;
int height = 250 * density;
//设置对话框尺寸
cd.setSize(width,height);
cd.setAutoClosable(true);
cd.setButton(0, "关闭", new IDialog.ClickedListener() {
    @Override
    public void onClick(IDialog iDialog, int i) {
        cd.destroy();
    }
});
//定制对话框
cd.setContentCustomComponent(dl);
cd.show();
```

这段代码使用了名为 custom_dialog.xml 的布局文件,该布局文件的代码如下:

```
<?xml version="1.0" encoding="utf-8"?>
<DirectionalLayout
    xmlns:ohos="http://schemas.huawei.com/res/ohos"
    ohos:height="match_content"
    ohos:width="match_parent"
    ohos:orientation="vertical"
    ohos:padding="5vp">
    <Text
        ohos:height="match_content"
        ohos:width="match_parent"
        ohos:text_color="#FF0000"
        ohos:text_size="25fp" ohos:text=" 这是标题 "
        ohos:text_alignment="center"/>
    <DirectionalLayout
        ohos:height="match_content"
```

```
            ohos:width="match_parent"
            ohos:orientation="horizontal"
            ohos:alignment="center">
        <Text
            ohos:height="match_content"
            ohos:width="match_parent"
            ohos:id="$+id:content"
            ohos:text_color="#0000FF"
            ohos:text_size="18fp"
            ohos:text_alignment="vertical_center"/>
    </DirectionalLayout>
</DirectionalLayout>
```

运行程序,会显示如图 6-6 所示的对话框。

图 6-6　定制对话框

6.3　Toast 信息框

HarmonyOS 还提供了一种 Toast 信息框。Toast 信息框与前面讲的对话框的主要区别是前者不会获得焦点,而且是非模态显示。也就是说,在 Toast 信息框显示时,仍然可以操作后面的组件,而且 Toast 信息框在过一段时间(可由用户指定关闭时间)后会自动关闭。而对话框是模态显示的,一旦显示了对话框,除非关闭对话框,否则无法操作对话框后面的组件。Toast 信息框一般用于向用户反馈一些信息,而又不影响用户操作其他组件的场景。

ToastDialog 类用于显示 Toast 信息框,通过 setComponent() 方法设置内容视图,通过 show() 方法显示 Toast 信息框。代码如下:

```
Text text = new Text(this);
text.setWidth(800);
text.setHeight(250);
text.setTextSize(100);
text.setText("这是一个信息框");
text.setPadding(30, 20, 30,20);
text.setMultipleLine(true);
text.setTextColor(Color.WHITE);
text.setTextAlignment(TextAlignment.CENTER);
ShapeElement style = new ShapeElement();
style.setShape(ShapeElement.RECTANGLE);
style.setRgbColor(new RgbColor(0x666666FF));
style.setCornerRadius(15);
text.setBackground(style);
DirectionalLayout mainLayout = new DirectionalLayout(this);
mainLayout.setWidth(800);
mainLayout.setHeight(250);
mainLayout.setAlignment(LayoutAlignment.CENTER);
mainLayout.addComponent(text);
ToastDialog td = new ToastDialog(this);
td.setSize(800, 250);
td.setDuration(300);          //3s 后消失
td.setTransparent(true);
td.setAlignment(LayoutAlignment.CENT ER);
td.setComponent(mainLayout);
td.show();
```

Toast 信息框一般是通过按钮单击事件触发的，将上述代码作为按钮的单击事件的响应代码，单击按钮后则会显示如图 6-7 所示的 Toast 信息框，在 3s 后，Toast 信息框会自动关闭。

图 6-7　信 息 框

6.4 Popup 对话框

气泡对话框是覆盖在当前界面之上的弹出框,可以相对组件或者屏幕显示。显示时会获取焦点,中断用户操作,被覆盖的其他组件无法交互。气泡对话框内容一般简单明了,并提示用户一些需要确认的信息。

PopupDialog 类用于显示气泡对话框,通过 PopupDialog()方法构造气泡对话框,setHasArrow()方法显示箭头,setArrowOffset()方法设置箭头的偏移量,setArrowSize()方法设置箭头的尺寸,setBackColor()方法设置气泡对话框的背景颜色,show()方法显示气泡对话框。代码如下:

```
PopupDialog pd = new PopupDialog(getContext(), component);
pd.setText("这是气泡对话框");
//显示箭头
pd.setHasArrow(true);
//更改箭头偏移量
pd.setArrowOffset(100);
//更改箭头尺寸
pd.setArrowSize(100,75);
//设置箭头背景颜色
pd.setBackColor(new Color(0xFFBEEDC7));
pd.show();
```

运行程序,单击"单击显示 PopupDialog",会显示如图 6-8 所示的 Popup 对话框。

图 6-8 Popup 对话框

第 7 章 多 媒 体

随着科技和移动设备的发展,超清摄像头和强大的成像算法使得目前智能手机的拍照功能具有非常好的效果。5G 时代的高速网络也使得传输大量音视频内容成为主流,并且内容的体量越来越大,人们对于手机的多媒体能力的要求也越来越高。本章主要讲解在 HarmonyOS 中如何开发多媒体的应用。

通过阅读本章,读者可以掌握:
➤ 如何进行音视频开发。
➤ 如何进行相机开发。
➤ 了解 HarmonyOS 相机的运行逻辑。

7.1 音　　频

音频作为最常见的多媒体展现形式,出现在各种各样的应用上面:提示音、铃声、音乐等。HarmonyOS 音频模块支持音频业务的开发,提供音频相关的功能,主要包括音频播放、音频采集、音量管理和短音播放等,本节主要涉及音频播放功能的开发。

7.1.1 准备本地音频文件

将预先准备好的音频文件放入项目结构的/entry/resources/base/media 目录中。如图 7-1 所示,本节所用资源为"sample.mp3"。

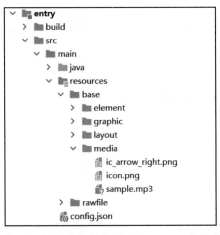

图 7-1　媒体文件的放置路径:media 文件夹

在 IDE 界面的最右侧找到 Gradle 工具,具体位置如图 7-2 所示。打开 Gradle 工具,依照如下路径:"项目名称"→"entry"→"ohos:debug"→"compileDebugResources",双击 compileDebugResources,如图 7-3 所示(compileDebugResources 用来打包调试用的资源文件,将外部静态资源打包成项目可用的资源文件)。

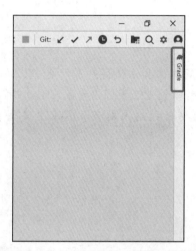

图 7-2 Gradle 工具在 IDE 界面上的位置   图 7-3 compileDebugResources 在 Gradle 的位置

Gradle 工具运行完毕后,在底边运行栏中会出现"[compileDebugResources]: successful"字样,如图 7-4 所示,说明音频已经被打包成开发者调试可用的资源文件。

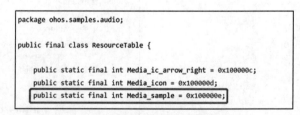

图 7-4 Gradle 工具[compileDebugResources]运行成功提示

开发者可以在项目中单击"entry"→"build"→"generated"→"source"→"r"→"项目名称"→"ResourceTable",找到已被打包的资源 ID,用来后续在 slice 中调用资源文件,如图 7-5 和图 7-6 所示。

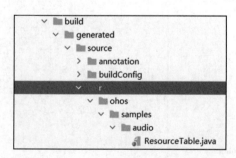
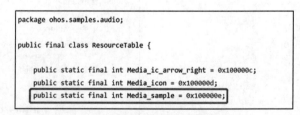

图 7-5 ResourceTable 的文件路径   图 7-6 ResourceTable 文件内部的结构

至此完成本地音频文件的准备工作。

接下来需要通过资源文件的 ID(也就是图 7-6 中方框圈出的部分)获取资源文件的路径,代码如下:

```
public String getPathById(Context context, int id) {
    String path = "";
    if(context == null) {
        return path;
    }
    ResourceManager manager = context.getResourceManager();
    if(manager == null) {
        return path;
    }
    path = manager.getMediaPath(id);
    return path;
}
```

这个函数可以通过输入的参数 ID 在 ResourceManager 中获取资源文件的路径。

7.1.2 播放本地音频文件

HarmonyOS 提供了功能强大的播放器 Player 类,支持当前各种主流音视频格式的播放。Player 提供了方便的音视频源接口,可以方便地读取并播放本地音视频和远程网络音视频,本节主要讲解如何通过 Player 类播放音频文件,视频文件的播放在 7.2 节给出。

要播放本地音频,先要完成 7.1.1 节的准备工作,读者可以随意准备一段音频,并按照 7.1.1 节中的步骤将其放到资源文件目录中。

对于本地音频,需要配置好 Player 对象的音频源,所以首先要生成音频文件的 BaseFileDescriptor 对象。与 7.1.1 节类似,需要将音频资源放到项目资源目录下,通过生成 ResourceID 来获取音频文件的路径,详见 7.1.1 节的 getPathById()函数。

使用该路径获取一个 RawFileEntry 实例对象,调用该实例对象的 openRawFileDescriptor() 方法获取 BaseFileDescriptor 对象,代码如下:

```
public BaseFileDescriptor getBFD(Context context, String path) {
    if(path == null || path.isEmpty()) {
        return null;
    }
    //通过资源文件路径获取 RawFileEntry 实例
    RawFileEntry assetManager = context.getResourceManager()
                        .getRawFileEntry(path);
    try {
        //通过 RawFileEntry 打开资源文件,也就是准备好的音频
        BaseFileDescriptor bfd = assetManager.openRawFileDescriptor();
        return bfd;
    } catch (IOException e) {
        e.printStackTrace();
    }
    return null;
}
```

通过使用 setSource()函数,为 Player 对象确定音频源。接下来只需调用 Player 类的

函数，即可对音频进行播放，代码如下：

```
String audioPath = getPathById(this, ResourceTable.Media_sample);
BaseFileDescriptor bfd = getBFD(this, audioPath);
Player player = new Player(this);
player.setSource(bfd);//为player添加音频源
player.prepare();
player.play();
```

7.1.3 暂停和继续播放音频

可以通过 Player 类中的 pause()成员函数来对音频暂停，play()函数继续播放。具体代码如下：

```
player.prepare();        //准备好媒体播放的环境和缓冲媒体数据
player.play();           //播放媒体
player.pause();          //暂停播放
player.play();           //继续播放
player.reset();          //从头开始播放
```

7.1.4 停止播放音频

当播放任务完成后，需要将分配给 Player 对象的资源释放掉，这里需要调用 release()函数释放资源。具体代码：

```
player.stop();
player.release();        //不需要再播放时，记得即时释放资源
```

7.1.5 播放在线音频文件

播放网络音频文件，其在播放环节与本地音频播放完全一致，差别仅在资源获取环节，Player 可以使用参数类型为 Source 类的 setSource()函数设置音频源，Source 类实例可以通过统一资源标识符(Uniform Resource Identifier, URI)来创建。如在上述例子中，只需将"为 Player 添加音频源"这一过程替换为如下两行代码：

```
Source source = newSource("http://.../在线音源.mp3");
player.setSource(source);
```

括号中的 URI 可以替换为自己的网络音频资源，播放的效果与播放本地音频的效果完全一致。

7.1.6 播放音频的完整案例

代码如下：

```
public class AudioTest {
    private Player player;
```

```java
        private String audioPath;
        private BaseFileDescriptor bfd;
        @Override
        public void onStart(Intent intent) {
            super.onStart(intent);
            audioPath = getPathById(this, ResourceTable.Media_sample);
            //读取本地资源
            bfd = getBFD(this, audioPath);
            player = new Player(this);              //实例化 Player 组件
            player.setSource(bfd);                  //为 Player 添加音频源
            player.prepare();                       //准备
            player.play();                          //播放
            player.pause();                         //暂停
            player.release();                       //释放
        }
        //获取资源文件路径
        public String getPathById(Context context, int id) {
            String path = "";
            if(context == null) {
                return path;
            }
            ResourceManager manager = context.getResourceManager();
            if(manager == null) {
                return path;
            }
            path = manager.getMediaPath(id);
            return path;
        }
        public BaseFileDescriptor getBFD(Context context, String path) {
            if(path == null || path.isEmpty()) {
                return null;
            }
            //通过资源文件路径获取 RawFileEntry 实例
            RawFileEntry assetManager = context.getResourceManager()
                .getRawFileEntry(path);
            try {
                //通过 RawFileEntry 打开资源文件，也就是音频文件
                BaseFileDescriptor bfd = assetManager.openRawFileDescriptor();
                return bfd;
            } catch (IOException e) {
                e.printStackTrace();
            }
            return null;
        }
}
```

7.2 视 频

随着移动终端的快速发展和普及，短视频内容获得了各大平台、用户的青睐，出现了像抖音、快手、微博、小红书等霸榜各大应用市场的杀手级应用，出现了大批的优秀短视频创作者。短视频平台已经不仅仅是娱乐的地方，还是各种信息传播的地方。HarmonyOS 视频模块支持视频业务的开发和生态开放，开发者可以通过已开放的接口很容易地实现视频媒体的播放、操作和新功能开发。视频媒体的常见操作有视频编解码、视频合成、视频提取、视频播放以及视频录制等，本节主要涉及视频播放功能的开发。

在 7.1.2 节中讲到 Player 类可以用来播放音频，实际上 Player 类同样可以实现视频的播放，所以对于视频（如 MP4 等格式）的播放可以参照 7.1.6 中的例子，与音频播放不同的是，播放视频需要配置实现一个 UI 组件来负责将视频播放出来。对于视频声音，Player 类可以自动连接到手机的音频播放组件，所以可以直接传出声音，而画面部分解码出来后却不知道将画面信息传递到哪里，因此必须手动创建一个 UI 组件，并将其和 Player 连接起来，用于显示由 Player 类对视频源解码得到的画面。在这里推荐使用 HarmonyOS 提供的 SurfaceProvider 组件，这个组件能够提供一个窗口，将画面信息显示于窗口上，下面介绍具体步骤：

（1）创建一个类，实现 SurfaceOps.Callback 接口。

（2）重写 SurfaceOps.Callback 中的方法，将回调函数 SurfaceOps() 添加到 Player 实例中。

（3）为 SurfaceProvider 实例中的 SurfaceOps() 添加回调。

具体代码如下：

```java
public class VideoTest extends AbilitySlice implements SurfaceOps
                        .Callback {
    private Player player;
    private SurfaceProvider surfaceProvider;
    private SurfaceOps surfaceOps;
    private DirectionalLayout myLayout;
    private DirectionalLayout.LayoutConfig layoutConfig;
    private String videoPath;
    private BaseFileDescriptor bfd;
    @Override
    public void onStart(Intent intent) {
        super.onStart(intent);
        this.setDisplayOrientation(AbilityInfo.DisplayOrientation
            .PORTRAIT);
        videoPath = getPathById(this, ResourceTable.Media_testpic1);
        bfd = getBFD(this, videoPath);           //读取本地视频
        player = new Player(this);               //实例化 Player 组件
        player.setSource(bfd);                   //为 Player 添加视频源
        //创建布局，用来承载 SurfaceProvider
        myLayout = new DirectionalLayout(this);
        //实例化 SurfaceProvider 组件，并为其设置属性
```

```
    surfaceProvider = new SurfaceProvider(this);
    layoutConfig = new DirectionalLayout.LayoutConfig(1080,607);
    layoutConfig.setMargins(0,0,0,0);
    surfaceProvider.setLayoutConfig(layoutConfig);
    //为SurfaceProvider 的 SurfaceOps 添加回调
    surfaceProvider.getSurfaceOps().get().addCallback(this);
    surfaceProvider.pinToZTop(true);
    //配置暂停按钮
    Button button_pause = new Button(this);
    button_pause.setLayoutConfig(layoutConfig);
    button_pause.setText("暂停");
    button_pause.setTextSize(100);
    button_pause.setMultipleLine(true);
    ShapeElement buttonBackground = new ShapeElement();
    buttonBackground.setRgbColor(new RgbColor(0x3c6c9c));
    buttonBackground.setCornerRadius(25);
    button_pause.setBackground(buttonBackground);
    //配置播放按钮
    Button button_play = new Button(this);
    button_play.setLayoutConfig(layoutConfig);
    button_play.setText("播放");
    button_play.setTextSize(100);
    button_play.setMultipleLine(true);
    buttonBackground.setRgbColor(new RgbColor(0x3c6c9c));
    buttonBackground.setCornerRadius(25);
    button_play.setBackground(buttonBackground);
    myLayout.addComponent(surfaceProvider);
    myLayout.addComponent(button_pause);
    myLayout.addComponent(button_play);
    button_pause.setClickedListener(new Component.ClickedListener(){
        @Override
        public void onClick(Component component) {
            player.pause();
        }
    });
    button_play.setClickedListener(new Component.ClickedListener() {
        @Override
        public void onClick(Component component) {
            player.prepare();
            player.play();
        }
    });
    super.setUIContent(myLayout);
}
//获取视频路径
public String getPathById(Context context, int id) {
    String path = "";
```

```java
        if(context == null) {
            return path;
        }
        ResourceManager manager = context.getResourceManager();
        if(manager == null) {
            return path;
        }
        path = manager.getMediaPath(id);
        return path;
    }
    public BaseFileDescriptor getBFD(Context context, String path) {
        if(path == null || path.isEmpty()) {
            return null;
        }
        //通过资源文件路径获取 RawFileEntry 实例
        RawFileEntry assetManager = context.getResourceManager()
                    .getRawFileEntry(path);
        try {
            //通过 RawFileEntry 打开资源文件,也就是视频文件
            BaseFileDescriptor bfd = assetManager.openRawFileDescriptor();
            return bfd;
        } catch (IOException e) {
            e.printStackTrace();
        }
        return null;
    }
    @Override
    public void surfaceCreated(SurfaceOps surfaceOps) {
        //将 Player 与 SurfaceOps 建立连接,也就是与 SurfaceProvider 建立了连接
        player.setSurfaceOps(surfaceOps);
        player.prepare();         //准备
        player.play();            //播放
    }
    @Override
    public void surfaceChanged(SurfaceOps surfaceOps, int i, int i1,
                    int i2) {}
    @Override
    public void surfaceDestroyed(SurfaceOps surfaceOps) {
        player.release();         //不需要再播放时,记得即时释放资源
    }
}
```

经过以上操作最终得到如图 7-7 所示的视频播放效果。

图 7-7 视频播放结果图

7.3 相　　机

随着科技的发展，相机已经是各大智能设备必备的一个硬件单元，相机的性能也渐渐成为了各个品牌智能设备的竞争因素。HarmonyOS 相机模块支持相机业务的开发，开发者可以通过已开放的接口实现相机硬件的访问、操作和新功能开发，最常见的操作如预览、拍照、连拍和录像等，本节主要涉及预览和拍照功能的开发。

7.3.1 拍照 API 的使用方式

在进行正式的开发之前，需要先了解相机开发的基本流程，如图 7-8 所示。还需要知道几个重要的 jar 包，如表 7-1 所示。

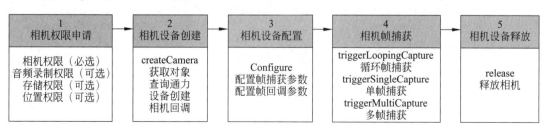

图 7-8 相机开发关键步骤

表 7-1 相机模块常用的 jar 包

包 名	说 明
ohos.media.camera.CameraKit	相机功能入口类。获取当前支持的相机列表及其静态能力信息,创建相机对象
ohos.media.camera.device	相机设备操作类。提供相机能力查询、相机配置、相机帧捕获、相机状态回调等功能
ohos.media.camera.params	相机参数类。提供相机属性、参数和操作结果的定义

7.3.2 使用相机需要申请的权限

首先需要在 config.json 文件中声明需要添加的权限。下面以相机权限及录音权限为例,展示如何添加这两个权限的信息,代码如下:

```
"reqPermissions": [
    {
        "name": "ohos.permission.CAMERA"
    },
    {
        "name": "ohos.permission.MICROPHONE"
    }
]
```

除了这两个权限之外,若需要保存图片至设备的外部,则需要申请存储权限 ohos.permission.WRITE_USER_STORAGE,若需要保存图片的位置信息,则需要申请位置权限 ohos.permission.LOCATION。

config.json 文件配置完成之后,就可以直接在设置中开启应用的对应权限或者直接让应用主动询问是否开启。

还可以对权限进行动态配置,同时可以查询目标权限是否开启,代码如下:

```
//主动询问权限是否开启
String[] permission={"ohos.permission.MICROPHONE",
                    "ohos.permission.CAMERA"};
for(int i=0;i<permission.length;i++){
    if(verifyCallingOrSelfPermission(permission[i]) != 0){
        if(canRequestPermission(permission[i])){
            requestPermissionsFromUser(permission, 0);
        }
    }
}
```

7.3.3 拍照的完整案例

1. 创建相机 UI

在 HarmonyOS 中实现相机的画面预览,可以通过创建 SurfaceProvider 实现。参照

7.2 节视频播放部分的知识,创建用于实现相机预览和照片展示的 SurfaceProvider 实例 surfaceProvider1 和 surfaceProvider2,随后建立一个用于拍照的 Button,将三者添加到布局中,具体代码如下:

```java
//定义 SurfaceProvider 的大小
public static final int VIDEO_WIDTH = 1080;
public static final int VIDEO_HEIGHT = 1080;
//删除导航栏等
Window window = this.getWindow();
window.setStatusBarVisibility(Component.INVISIBLE);
window.setNavigationBarColor(Color.TRANSPARENT.getValue());
//布局容器
DirectionalLayout myLayout = new DirectionalLayout(this);
DirectionalLayout.LayoutConfig config = new DirectionalLayout
        .LayoutConfig(DirectionalLayout.LayoutConfig.MATCH_PARENT,
        DirectionalLayout.LayoutConfig.MATCH_PARENT);

myLayout.setLayoutConfig(config);
myLayout.setOrientation(Component.HORIZONTAL);
ShapeElement background = new ShapeElement();
background.setRgbColor(new RgbColor(255,255,255));
myLayout.setBackground(background);
//SurfaceProvider,用于进行摄像头画面预览
config.width = VIDEO_WIDTH;
config.height = VIDEO_HEIGHT;
config.setMargins(0,0,0,0);
config.alignment = LayoutAlignment.VERTICAL_CENTER;
surfaceProvider = new SurfaceProvider(this);
surfaceProvider.setLayoutConfig(config);
surfaceProvider.getSurfaceOps().get().addCallback(callback);
surfaceProvider.pinToZTop(true);
//Button,单击拍摄照片
config.height = DirectionalLayout.LayoutConfig.MATCH_CONTENT;
config.width = DirectionalLayout.LayoutConfig.MATCH_CONTENT;
config.setMargins(450,20,0,0);
Button button = new Button(this);
button.setLayoutConfig(config);
button.setText("拍照");
button.setTextSize(100);
button.setMultipleLine(true);
ShapeElement buttonBackground = new ShapeElement();
buttonBackground.setRgbColor(new RgbColor(0x3c6c9c));
buttonBackground.setCornerRadius(25);
button.setBackground(buttonBackground);
button.setClickedListener(new Component.ClickedListener() {
    @Override
    public void onClick(Component component) {
        cameraDevice.triggerSingleCapture(takePictureConfig);        //拍照
    }
});
```

```java
//添加组件到布局容器中,并将布局容器作为 UI 的根布局
myLayout.addComponent(surfaceProvider);
myLayout.addComponent(button);
super.setUIContent(myLayout);
```

在这段代码中定义了一个宽 1080,高 1080 的预览窗口,并且增加了描述 surfaceProvider 的回调 callback,该回调将定义该组件在其生命周期不同阶段的执行内容,callback 的具体代码如下:

```java
//callback 的具体代码
private SurfaceOps.Callback callback = new SurfaceOps.Callback() {
    @Override
    public void surfaceCreated(SurfaceOps surfaceOps) {
        //将 previewSurface 作为预览相机画面
        previewSurface = surfaceOps.getSurface();
        openCamera();            //打开相机
    }
    @Override
    public void surfaceChanged(SurfaceOps surfaceOps,int i, int i1,
    int i2) {}
    @Override
    public void surfaceDestroyed(SurfaceOps surfaceOps) {}
};
```

在这段代码中,使用 surfaceOps.getSurface()方法获取 surfaceProvider,用于预览显示 Surface 的对象 previewSurface,并且用 openCamera()方法来创建相机对象。openCamera()方法中创建相机对象的方法参考下一步。

以上代码创建的相机 UI 效果如图 7-9 所示。

图 7-9　用于展示相机能力的 UI 界面

从上至下分别是用于预览相机画面 surfaceProvider 和拍照按钮 Button。

2. 创建相机对象

在 HarmonyOS 上实现一个相机应用,无论要应用在哪些设备上,都必须先创建一个相机设备,才能进行后序操作。CameraKit 类,是相机的入口 API 类,用于获取相机设备特性、打开相机等。在 HarmonyOS 中,相机设备的创建首先需要创建一个 CameraKit 实例并通过 CameraKit.getInstance(Context context)方法获取唯一的 CameraKit 对象,代码如下:

```
//获取 CameraKit 对象
private void openCamera(){
    cameraKit=CameraKit.getInstance(this);
    if(cameraKit == null) {
        //处理 cameraKit 获取失败的情况
        return;
    }
}
```

如上述代码所示,在这里可以使用 getInstance()方法获取 CameraKit 对象。如果此步骤操作失败,相机可能已经被占用或无法使用。如果被占用,则必须等到相机被释放后才能重新获取 CameraKit 对象。接下来需要获取设备支持的逻辑相机列表,并基于某个逻辑相机创建相机对象,具体实现这一方法需要以下 4 个步骤。

(1) 通过调用 getCameraIds()函数,获取当前设备可用的逻辑相机列表。此列表存储了当前设备所有可用的逻辑相机 ID,如果列表不为空,则可以在列表中选择其中一个 ID 用来创建相机对象,否则,表示当前设备没有可用的相机,则不能进行下一步。

(2) 通过创建 CameraStateCallback 的实例对象来为相机对象添加回调函数,实现 CameraStateCallbackImpl()方法创建用于接收有关摄像机设备的状态更新的回调对象,必须提供此回调实例才能打开相机设备。

(3) 创建执行回调的 EventHandler。

(4) 通过使用 HarmonyOS 提供的 createCamera(String camerald,CameraStateCallback callback,EventHandler handler)方法来创建一个可用的相机对象,如果此步骤执行成功,则表示相机系统的硬件已经自动完成了上电的操作。当通过该方法完成相机创建之后,会在 CameraStateCallback 中触发 onCreated(Camera camera)回调。在进入相机设备的配置步骤前,需确保相机设备已经创建成功。

以上 4 个步骤可以通过下面的代码实现,将该部分的代码放在获取 CameraKit 对象之后运行即可完成完整的相机创建流程,代码如下:

```
try {
    //配置 imageReceiver,用于接收拍摄的画面帧
    imageReceiver=ImageReceiver.create(SCREEN_HEIGHT,SCREEN_WIDTH,
                ImageFormat.JPEG,IMAGE_RCV_CAPACITY);
    //获取当前设备的逻辑相机列表 cameraIds
    String[] cameraIds = cameraKit.getCameraIds();
```

```
    //用于相机创建和相机运行的回调
    CameraStateCallbackImpl cameraStateCallback = new
    CameraStateCallbackImpl ();
    //创建用于运行相机的线程
    eventHandler = new EventHandler(EventRunner.create("CameraCb"));
    //创建相机
    cameraKit.createCamera(cameraIds[0],cameraStateCallback,eventHandler);
} catch (IllegalStateException e) {
    System.out.println("error:fail to get cameraIds");
}
```

在相机创建的同时可以获取与相机相关的信息,例如,调用 getDeviceLinkType(String physicalId)方法可以获取物理相机的连接方式信息;调用 getCameraInfo(String cameraId)方法可以查询相机硬件朝向等信息;通过调用 getCameraAbility(String cameraId)方法可以查询相机能力信息(如支持的分辨率列表等)。更多的详细方法和调用信息可以通过官方的 API 文档进行查询。

3. 相机配置

前文讲到相机创建后会在 CameraStateCallback 中触发 onCreated(Camera camera)回调,下面介绍 CameraStateCallback 中的函数,代码如下:

```
//CameraStateCallbackImpl
private final class CameraStateCallbackImpl extends CameraStateCallback {
    //相机回调
    @Override
    public void onCreated(Camera camera) {
        //相机创建时回调
        previewSurface = surfaceProvider.getSurfaceOps().get().getSurface();
        CameraConfig.Builder cameraConfigBuilder = camera.
                                                getCameraConfigBuilder();
        //配置预览的 Surface
        cameraConfigBuilder.addSurface(previewSurface);
        //配置拍照的 Surface
        cameraConfigBuilder.addSurface(imageReceiver
                                .getRecevingSurface());
        try {
            //相机设备配置
            camera.configure(cameraConfigBuilder.build());
            cameraDevice = camera;
        } catch (IllegalArgumentException e) {
            System.out.println("Argument Exception");
        } catch (IllegalStateException e) {
            System.out.println("State Exception");
        }
    }
    @Override
```

```
    public void onConfigured(Camera camera) {
        //相机配置
        FrameConfig.Builder frameConfigBuilder = cameraDevice
            .getFrameConfigBuilder(FRAME_CONFIG_PREVIEW);
        //配置预览 Surface
        frameConfigBuilder.addSurface(previewSurface);
        //启动循环帧捕获
        camera.triggerLoopingCapture(frameConfigBuilder.build());
    }
    @Override
    public void onReleased(Camera camera) {
        //释放相机设备
        if(camera != null) {
            camera.release();
            camera = null;
        }
    }
}
```

4. 相机拍照

为了能在画面中显示相机捕获到的一帧画面，在 onCreated(Camera camera)方法中为 CameraConfig 实例额外配置了一个 imageReceiverSurface，在 onConfigured(Camera camera)函数中创建了一个 FrameConfig 实例，选用了预览模式并配置了预览 Surface 实例 previewSurface。随后启动循环帧捕获，相机拍摄的每一帧画面都会传输到 previewSurface 上，从而实现了相机画面的预览。每按一次按钮就取一帧画面到 imageReceiverSurface，从而实现拍照画面的展示效果，按钮的逻辑代码如下：

```
button.setClickedListener(new Component.ClickedListener() {
    @Override
    public void onClick(Component component) {
        //配置拍照设置
        FrameConfig.Builder framePictureConfigBuilder = cameraDevice
                .getFrameConfigBuilder(FRAME_CONFIG_PICTURE);
        framePictureConfigBuilder.addSurface(imageReceiver
                .getRecevingSurface());
        FrameConfig pictureFrameConfig=framePictureConfigBuilder.build();
        cameraDevice.triggerSingleCapture(pictureFrameConfig); //拍照
    }
});
```

每当按钮被触发就会回调 onClick(Component component)，在其中配置 pictureFrameConfig 并把 imageReceiver 添加进去，这样拍摄的照片就会通过 imageReceiver 显示在容器上，效果如图 7-10 所示。

当按下拍摄按钮之前，画面是动态的，按下拍摄按钮之后，画面变为静态。

图 7-10 实现预览拍照后的 UI 界面

5. 切换镜头

目前市面上的智能移动终端，基本上都配备了多摄像头，主要分布为后置摄像头和前置摄像头，当然也有相当多的机型在前置和后置都配备了多个不同用途的镜头。接下来主要讲解切换摄像头的步骤。

（1）查询镜头类型。

上文提到过，打开相机需要先查询可用的 cameraIds，再利用逻辑相机 ID 去打开对应的镜头。这里可以利用实例化的 cameraKit 对象去查询 cameraId 属于什么类型的镜头。通过静态调用 CameraInfo 类中的 FacingType 接口，对比对应的镜头类型，并且用一个全局变量 isFront（boolean 类型）表示当前摄像头是否为前置摄像头。将以下代码添加到 openCamera() 函数中：

```
String cameraId = "";
for(String logicalCameraId: cameraIds){
    int cameraType = cameraKit.getCameraInfo(logicalCameraId).getFacingType();
    switch (cameraType){
        case CameraInfo.FacingType.CAMERA_FACING_FRONT:
            if(isFront) {
                cameraId = logicalCameraId;
            }
        case CameraInfo.FacingType.CAMERA_FACING_BACK:
            if(!isFront) {
                cameraId = logicalCameraId;
            }
            break;
```

```
        }
    }
    cameraKit.createCamera(cameraId, cameraStateCallback, eventHandler);
```

（2）添加切换按钮。

在布局中添加切换按钮，创建一个按钮对其进行配置，并设置一个 Click 监听事件。代码如下：

```
Button button_switch = new Button(this);
button_switch.setLayoutConfig(config);
button_switch.setText("切换");
button_switch.setTextSize(100);
button_switch.setMultipleLine(true);
buttonBackground.setRgbColor(new RgbColor(0x3c6c9c));
buttonBackground.setCornerRadius(25);
button_switch.setBackground(buttonBackground);
myLayout.addComponent(button_switch);
button_switch.setClickedListener(new Component.ClickedListener() {
    @Override
    public void onClick(Component component) {
        isFront = !isFront;            //isFront 取反，表示切换镜头类型
        if(cameraDevice!=null){
            cameraDevice.release();    //释放相机资源
        }
        openCamera();                  //重新打开相机
    }
});
```

添加了以上代码，就可以通过单击"切换"按钮对相机进行切换，效果如图 7-11 所示。

图 7-11　切换镜头效果

第 8 章 数据管理

数据是一个应用非常重要的组成部分,保存应用的配置、存储用户信息、检索英文单词,都需要用到数据功能。HarmonyOS 不仅可以操作本地数据,还可以对分布式数据和分布式文件进行操作,这是 HarmonyOS 独有的特性。在本章中将会对 HarmonyOS 操作本地数据、分布式文件和分布式数据的 API 的使用方法进行详细介绍。

通过阅读本章,读者可以掌握:
- 轻量级存储开发。
- 关系型数据库 SQLite 开发使用。
- 对象关系映射数据库开发。
- 分布式文件开发。
- 分布式数据开发。

8.1 轻量级数据存储开发

大多数应用都会或多或少使用一些文件来保存应用的配置信息,如用户名、账号密码、用户操作的最后状态等。这些文件通常使用文本格式,最常用的有 JSON 和 XML。轻量级数据存储功能通常可以用于保存应用的一些常用配置信息,并不适合需要存储大量数据和频繁改变数据的场景,但是用来保存少量数据非常合适。应用的数据被保存在文件中,这些文件会被持久化地存储在设备上。

8.1.1 Preferences 类的基本用法

Preferences 类用于读写纯文本格式的缓存文件,和安卓中的 SharedPreferences 类似。在纯文本格式的缓存文件中,数据是以 key-value 对的形式放在文件中,可以像操作 Map 一样操作这些缓存文件中的数据。

要获取 Preferences 对象,首先需要创建 DatabaseHelper 对象,然后通过 getPreferences()方法获取 Preferences 对象,创建代码如下:

```
DatabaseHelper infoDatabaseHelper = new DatabaseHelper(context);
Preferences infoPreferences = infoDatabaseHelper.getPreferences("info");
```

info 是存取数据文件的名字。这里不需要指定路径,因为系统会给 info 文件指定默认的存储路径。如果要获取 info 文件的路径,可以使用 context.getPreferencesDir()方法,一般是/data/data/{PackageName}/{AbilityName}/preferences。

Preferences 类提供了一系列 putXxx() 方法和 getXxx() 方法,用来对文件中的 key-value 对进行读写,其中 Xxx 代表 String、Int、Float 等数据类型。通过执行 flush() 方法,可以将缓存的数据再次写回文本文件中进行持久化存储。Preferences 提供两种持久化的方式,flush() 是同步刷新的方式,flushSync() 是异步刷新的方式。

【例 8.1】 将学生信息保存在 StudentInfo.cfg 文件中,然后将学生信息输出到 HiLog 视图中。

```
public class PreferenceTestDemo {
    //将学生信息写入文件中
    public static void writeStudentInfo(Context context) {
        DatabaseHelper databaseHelper = new DatabaseHelper(context);
        //context 入参类型为 ohos.app.Context
        String fileName = "StudentInfo.cfg";    //文件名
        //获取 Preferences 对象
        Preferences preferences = databaseHelper.getPreferences(fileName);
        //存入学号
        preferences.putString("stuId", "21120345");
        //存入学生名字
        preferences.putString("stuName", "Jack");
        //存入学生年龄
        preferences.putInt("stuName", 19);
        //存入学生学习的课程
        Set<String> courses = new HashSet<>();
        courses.add("Chinese");
        courses.add("English");
        preferences.putStringSet("course", courses);
        preferences.flush();          //异步保存 data.cfg 文件
        preferences.flushSync();      //同步保存 data.cfg 文件
    }
    //读取学生信息
    public static void readStudentInfo(Context context) {
        DatabaseHelper databaseHelper = new DatabaseHelper(context);
        String fileName = "StudentInfo.cfg";
        Preferences preferences = databaseHelper.getPreferences(fileName);
        //开始读取数据,并将读取的数据输出到 HiLog 视图中
        //getXxx() 方法一般需要一个默认值,当没有获取到值的时候输出默认值
        HiLogs.info("学生学号:" + preferences.getString("stuId",
                                                    "defaultValue"));
        HiLogs.info("学生姓名:" + preferences.getString("stuName", null));
        HiLogs.info("学生年龄:" + preferences.getInt("stuAge", 20));
        HiLogs.info("学生学习的课程:" + preferences.getStringSet("course",
                                                        null));
        //获取文件的存储路径
        HiLogs.info("获取默认文件路径:" + context.getPreferencesDir().toString());
    }
}
```

写一个按钮的单击事件，依次调用写入学生信息方法和读取学生信息方法。输出结果如图 8-1 所示。

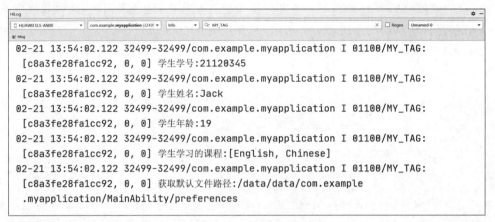

图 8-1　HiLog 视图中的输出

在控制台中输出了学生信息，并且获取了缓存文件放置的默认位置。按照路径找到文件打开以后可以查看其中数据的内容。

8.1.2　响应文件写入动作

Preferences 对象可以对文件的写入动作进行响应，响应写入动作的对象被称作观察者（Observer）。观察者对象对应的类必须实现 Preferences.PreferencesObserver 接口，该接口中的 onChange() 方法用来响应 key 的写入动作。订阅的 key 的值发生变更后，如在执行 flush() 方法时，PreferencesObserver 被触发回调，这时就会调用 onChange() 方法，该方法的原型如下：

```
public void onChange(Preferences preferences, String key);
```

其中，preferences 是要监听的 Preferences 对象，key 是已经写入的 key-value 对中的 key。观察者要被注册到 preferences 当中，即调用 preferences.registerObserver() 方法，方法原型如下：

```
public void registerObserver(PreferencesObserver preferencesObserver);
```

preferencesObserver 表示观察者的名字。在观察者使用结束以后，系统提供了注销观察者的方法，可以使用 preferences.unRegisterObserver() 方法注销观察者。

【例 8.2】　向 score 文件多次写入学生的分数 score，并通过观察者对象输出所有输入分数的最大值。

```
//学生英语成绩类
public class EnglishScore {
    //学生姓名
    private String stuName;
    //学号
```

```java
        private String stuId;
        //分数
        private int score;
        public EnglishScore(String stuName, String stuId, int score) {
            this.stuName = stuName;
            this.stuId = stuId;
            this.score = score;
        }
        //getter()和setter()方法用工具生成
}
public class PreferenceObserverDemo {
    //观察者对象
    private static class ScoreMaxObserver implements
    Preferences.PreferencesObserver {
        //分数最高的学生信息要用ZSONObject对象接入
        private ZSONObject maxStudent = null;
        @Override
        public void onChange(Preferences preferences, String key) {
            //其他的key写入的时候也会触发
            //只对存入的值的key是score的value进行操作
            if("score".equals(key)) {
                HiLogs.info("响应写入接口,开始计算最高分");
                String s_score = preferences.getString(key, null);
                ZSONObject zsonObject = ZSONObject.stringToZSON(s_score);
                int score = zsonObject.getInteger("score");
                if(maxStudent == null)
                    maxStudent = zsonObject;
                if(score > maxStudent.getInteger("score"))
                    maxStudent = zsonObject;
                //输出最高分学生信息
                HiLogs.info("目前最高分学生是: " +
                        maxStudent.getString("stuName")+",学号是: "+
                        maxStudent.getString("stuId ")+",最高分: "+
                        maxStudent.getInteger("score"));
            }
        }
    }
    //注册和注销观察者对象
    public static void scoreObserver(Context context) {
        DatabaseHelper databaseHelper = new DatabaseHelper(context);
        String fileName = "price.cfg";
        Preferences preferences = databaseHelper.getPreferences(fileName);
        //创建观察者对象
        ScoreMaxObserver scoreMaxObserver = new ScoreMaxObserver();
        //注册观察者对象
        preferences.registerObserver(scoreMaxObserver);
        //写入学生分数和学生信息
```

```
        EnglishScore bobScore = new EnglishScore("Bob","2112",85);
        EnglishScore jackScore = new EnglishScore("Jack","2113",95);
        EnglishScore mikeScore = new EnglishScore("Mike","2116",98);
        ZSONObject bobObject = ZSONObject.classToZSON(bobScore);
        ZSONObject jackObject = ZSONObject.classToZSON(jackScore);
        ZSONObject mikeObject = ZSONObject.classToZSON(mikeScore);
        //对象要转化成 JSON 格式才能存入
        preferences.putString("score", bobObject.toString());
        preferences.flushSync();
        preferences.putString("score", jackObject.toString());
        preferences.flushSync();
        preferences.putString("score", mikeObject.toString());
        preferences.flushSync();
        preferences.unregisterObserver(scoreMaxObserver);
    }
}
```

最后，在按钮的单击事件中调用 scoreObserver()方法。

运行程序，会在 HiLog 视图中输出如图 8-2 所示的内容。

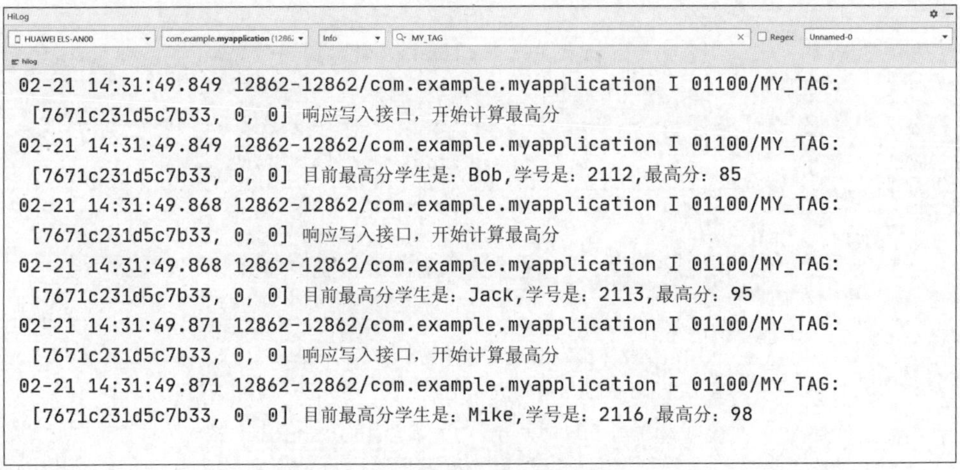
HiLog
HUAWEI ELS-AN00
com.example.myapplication (1286
Info
MY_TAG
Regex
Unnamed-0
hilog
02-21 14:31:49.849 12862-12862/com.example.myapplication I 01100/MY_TAG:
[7671c231d5c7b33, 0, 0] 响应写入接口，开始计算最高分
02-21 14:31:49.849 12862-12862/com.example.myapplication I 01100/MY_TAG:
[7671c231d5c7b33, 0, 0] 目前最高分学生是：Bob,学号是：2112,最高分：85
02-21 14:31:49.868 12862-12862/com.example.myapplication I 01100/MY_TAG:
[7671c231d5c7b33, 0, 0] 响应写入接口，开始计算最高分
02-21 14:31:49.868 12862-12862/com.example.myapplication I 01100/MY_TAG:
[7671c231d5c7b33, 0, 0] 目前最高分学生是：Jack,学号是：2113,最高分：95
02-21 14:31:49.871 12862-12862/com.example.myapplication I 01100/MY_TAG:
[7671c231d5c7b33, 0, 0] 响应写入接口，开始计算最高分
02-21 14:31:49.871 12862-12862/com.example.myapplication I 01100/MY_TAG:
[7671c231d5c7b33, 0, 0] 目前最高分学生是：Mike,学号是：2116,最高分：98

图 8-2　HiLog 视图中的输出

从控制台输出信息可以看到，本例输出了 3 次学生信息，每次都是由于调用 flushSync()方法后触发了观察者对象的 onChange()方法才输出这些信息的。

8.1.3　轻量级存储的移动和删除

Preferences 类提供了 movePreferences()来移动文件。movePreferences()方法的原型如下：

```
public boolean movePreferences(Context sourceContext, String sourceName,
                                String targetName);
```

其中，sourceName 和 targetName 分别表示原来的文件名和移动后的文件名。这个功能相当于重命名文件。该方法如果成功执行,即重命名成功,就会返回 true;否则,返回 false。

Preferences 类提供了 deletePreferences()方法来删除文件,其原型如下:

```
public boolean deletePreferences(String name);
```

name 参数表示要删除的文件名。如果文件被成功删除,返回 true;否则,返回 false。

要注意的是,deletePreferences()方法不仅会将 Preferences 实例从内存中移除,与此同时还会删除其在设备上的持久化文件。如果仅仅将 Preferences 实例从内存中移除,可以使用 removePreferencesFromCache()方法,其原型如下:

```
public void removePreferencesFromCache(String name);
```

【例 8.3】 新建一个学生信息类 userInfo 将其重命名为 userInfo2,然后将重命名之后的文件删除。

```
public class PreferencesDemo{
    //移动文件(重命名文件)
    public static boolean moveFile(Context context) {
        DatabaseHelper databaseHelper = new DatabaseHelper(context);
        String srcFile = "userInfo";
        Preferences preferences = databaseHelper.getPreferences(srcFile);
        preferences.putInt("age", 18);
        preferences.flush();
        String targetFile ="userInfo2";
        return databaseHelper.movePreferences(context, srcFile, targetFile);
    }
    //删除文件
    public static boolean deleteFile(Context context) {
        DatabaseHelper databaseHelper = new DatabaseHelper(context);
        String filename ="userInfo2";
        Preferences preferences =databaseHelper.getPreferences(filename);
        return databaseHelper.deletePreferences(filename);
    }
}
```

运行程序,单击"移动文件"按钮,会看到在系统默认文件路径下出现了一个 userInfo2 文件,然后再单击"删除文件"按钮,会看到 userInfo2 文件被删除了。

8.2 关系数据库开发

关系数据库是在 SQLite 数据库基础上实现的本地数据操作机制,提供给用户无须编写原生 SQL 语句就能进行数据增删改查的方法,同时也支持原生 SQL 语句操作,功能十分灵活。SQLite 数据库是移动端应用最常用的本地数据库。HarmonyOS 和 Android 一样,也提供了丰富的 API 用来操作 SQLite 数据库。本节将讲解如何使用 SQL 语句方式和使用谓词的方式来操作 SQLite 数据库。

8.2.1 使用 SQL 语句操作 SQLite 数据库

使用 SQL 是操作数据库最普遍也最简单的方式。HarmonyOS 提供了一系列 API 来执行 SQL 语句,其主要操作步骤如下。

(1) 通过 StoreConfig 对象指定数据库文件的路径、数据库名称、存储模式、是否只读等配置。

(2) 创建 DatabaseHelper 对象。DatabaseHelper 对象是数据库操作的辅助类,当数据库创建成功后,数据库文件将存储在上下文指定的目录里。

(3) RdbOpenCallback 在数据库创建时被回调。创建时必须实现 onCreate() 方法和 onUpgrade() 方法。onCreate() 方法会在第一次创建数据库时调用,通常在 onCreate() 方法中为数据库创建表(table)和视图(view);onUpgrade() 方法升级数据库时被调用。两个方法原型如下:

```
public abstract void onCreate(RdbStore store);
public abstract void onUpgrade(RdbStore store, int currentVersion,
                    int targetVersion);
```

还有一个 onDowngrade() 方法是在数据库降级时被调用,一般情况下不会用到,其原型如下:

```
public void onDowngrade(RdbStore store, int currentVersion,
                    int targetVersion);
```

(4) 获取 RdbStore 对象。RdbStore 主要作用就是操作数据库。该对象通过 DatabaseHelper 对象的 getRdbStore() 方法获取。

(5) 使用 executeSql() 方法负责执行非查询类的原生 SQL 语句,querySql() 方法负责执行原生的用于查询操作的 SQL 语句,并且返回 ResultSet 结果集对象。

【例 8.4】 创建一个学生信息数据库 student.sqlite 来对学生信息进行管理。该数据库包含一个名为 studentInfo 的表,表中包含学号、姓名和所在班级三个字段。依次对该表插入数据以及查询数据,最后将查询结果输出到 HiLog 视图中。

```java
public class SQLiteTestDemo {
    public static void sqliteWithSQL(Context context) {
        RdbStore store;
        //指定数据库名称
        StoreConfig config = StoreConfig.newDefaultConfig("student.sqlite");
        RdbOpenCallback callback = new RdbOpenCallback() {
            //创建表时调用
            @Override
            public void onCreate(RdbStore store) {
                //创建 studentInfo 表
                store.executeSql("CREATE TABLE IF NOT EXISTS studentInfo
                    (stuId INTEGER PRIMARY KEY, stuName VARCHAR(20),
                                    stuClass VARCHAR(20))");
```

```
            HiLogs.info("学生信息表创建成功");
        }
        @Override
        public void onUpgrade(RdbStore store, int oldVersion,
                                            int newVersion) {}
    };
    DatabaseHelper helper = new DatabaseHelper(context);
    store = helper.getRdbStore(config, 1, callback, null);
    String insertSQL ="insert into studentInfo
                            (stuId,stuName,stuClass)values(?,?,?);";
    //向学生信息表中插入4条记录,用Object数组方式进行插入
    store.executeSql(insertSQL, new Object[]{2112, "Mike", "计算机1班"});
    store.executeSql(insertSQL, new Object[]{2115, "Jay", "软工2班"});
    store.executeSql(insertSQL, new Object[]{2116, "Lee", "计算机2班"});
    store.executeSql(insertSQL, new Object[]{2117, "Mike", "计算机2班"});
    String selectSQL = "select * from studentInfo where stuName=?";
    //查询学生信息表中姓名Mike的学生信息
    ResultSet resultSet = store.querySql(selectSQL,
            new String[]{String.valueOf("Mike")});
    //有多个姓名为Mike的学生,遍历输出
    while(resultSet.goToNextRow()) {
    //输出查询结果
    HiLogs.info("学生学号:" + resultSet.getInt(0) + ",学生姓名:"
                        + resultSet.getString(1) + ",学生所在班级:"
                        + resultSet.getString(2));
    }
    store.close();//关闭数据库
}
```

单击"使用SQL操作SQLite数据库"按钮,会看到HiLog视图中输出如图8-3所示的查询结果。

图 8-3 输出查询结果

阅读这段代码,需要注意以下几点。

(1) 用StoreConfig对象指定数据库文件名时,不需要指定数据库文件路径,路径是固定的。

（2）RdbOpenCallback 对象中的 onUpgrade()方法在数据库版本升级后被调用，通常用于升级表结构，修改视图，或升级表中的数据。

（3）使用 querySql()方法返回的 ResultSet 对象的指针指向的是列表的表头，并不是第 1 条记录，有多条结果时可以使用 goToNextRow()全部进行遍历。

（4）ResultSet 对象提供了一系列 getXxx()方法，用于获取当前记录中特定字段的值，其中 Xxx 表示 Int、String、Double、Float 等。getXxx()方法的参数是字段索引，从 0 开始。

（5）使用 context.getDatabaseDir()方法可以获取数据库的存储目录，不过获取的是 databases 文件夹的目录。SQLite 数据库文件存放在 databases 文件夹下的 db 目录中，所以 student.sqlite 文件的完整路径如下：

```
/data/data/{packageName}/{ablityName}/databases/db/student.sqlite
```

8.2.2 使用谓词操作 SQLite 数据库

RdbStore 对象提供了 4 个方法：insert()、delete()、update()和 query()，分别用来对数据表进行增删改查操作。使用这 4 个方法操作数据库，不需要使用 SQL 语句。这 4 个方法的原型如下：

```
long insert(String table, ValuesBucket initialValues);
```

该方法通过返回值判断是否插入成功，插入成功时返回最新插入数据所在的行号，失败时则返回-1。

```
int delete(AbsRdbPredicates predicates);
```

该方法的返回值表示删除的数据行数，可根据此值判断是否删除成功。如果删除失败，则返回 0。

```
int update(ValuesBucket values, AbsRdbPredicates predicates);
```

该方法的返回值表示更新操作影响的行数。如果更新失败，则返回 0。

```
ResultSet query(AbsRdbPredicates predicates, String[] columns);
```

这 4 个方法中并不需要写 SQL 语句，例如，insert()方法的第 1 个参数 table 用于传递要插入数据的表名。第 2 个参数 ValuesBucket 存储的待插入的数据，它提供一系列 putXxx()方法，用于向 ValuesBucket 中添加数据，Xxx 表示 Int、String、Float 等数据类型。所有 ValuesBucket 类型的参数值传递的都是 key-value 对。

在这 4 个方法中还有 AbsRdbPredicates 类型的参数，表示谓词，指通过方法来描述值的大小和值的关系。AbsRdbPredicates 的实现类有两个：RdbPredicates 和 RawRdbPredicates。RdbPredicates 支持调用谓词提供的 equalTo 等接口，设置查询条件。RawRdbPredicates 仅支持设置表名、where 条件子句和 whereArgs 三个参数，不支持 equalTo 等接口调用。两个类常用接口如表 8-1 所示。

表 8-1　RdbPredicates 和 RawRdbPredicates 常用接口

类　　名	接　口　名	描　　述
RdbPredicates	RdbPredicates equalTo(String field,String value)	设置谓词条件,满足 field 字段与 value 值相等
RdbPredicates	RdbPredicates notEqualTo(String field, String value)	设置谓词条件,满足 field 字段与 value 值不相等
RdbPredicates	RdbPredicates beginsWith(String field, String value)	设置谓词条件,满足 field 字段以 value 值开头
RdbPredicates	RdbPredicates between(String field, int low, int high)	设置谓词条件,满足 field 字段在最小值 low 和最大值 high 之间
RdbPredicates	RdbPredicates orderByAsc(String field)	设置谓词条件,根据 field 字段升序排列
RawRdbPredicates	void setWhereClause(String whereClause)	设置 where 条件子句
RawRdbPredicates	void setWhereArgs(List<String> whereArgs)	设置 whereArgs 参数,该值表示 where 子句中占位符的值

对于操作结果集 ResultSet,HarmonyOS 也提供了一系列函数进行遍历和访问,如表 8-2 所示。

表 8-2　ResultSet 常用接口

接　口　名	描　　述
boolean goTo(int offset)	从结果集当前位置移动指定偏移量
boolean goToRow(int position)	将结果集移动到指定位置
boolean goToNextRow()	将结果集向后移动一行
boolean goToPreviousRow()	将结果集向前移动一行
boolean isStarted()	判断结果集是否被移动过
boolean isEnded()	判断结果集当前位置是否在最后一行之后
boolean isAtFirstRow()	判断结果集当前位置是否在第一行
boolean isAtLastRow()	判断结果集当前位置是否在最后一行
int getRowCount()	获取当前结果集中的记录条数
int getColumnCount()	获取结果集中的列数
String getString(int columnIndex)	获取当前行指定列的值,以 String 类型返回
byte[] getBlob(int columnIndex)	获取当前行指定列的值,以字节数组形式返回
double getDouble(int columnIndex)	获取当前行指定列的值,以 double 型返回

【例 8.5】使用谓词方式查询学生信息。

```
public class SQLiteDemo {
    //使用谓词操作 SQLite 数据库
```

```java
public static void sqliteWithPredicates(Context context) {
    RdbStore store;
    StoreConfig config = StoreConfig
            .newDefaultConfig("studentInfo.sqlite");
    RdbOpenCallback callback = new RdbOpenCallback() {
        //创建表时调用
        @Override
        public void onCreate(RdbStore store) {
            //创建 studentInfo 表
            store.executeSql("CREATE TABLE IF NOT EXISTS studentInfo(" +
                    "stuId INTEGER PRIMARY KEY, stuName VARCHAR(20)," +
                    "stuClass VARCHAR(20))");
            HiLogs.info("学生信息表创建成功");
        }
        public void onUpgrade(RdbStore store,
                              int oldVersion, int newVersion) {}
    };
    DatabaseHelper helper = new DatabaseHelper(context);
    store = helper.getRdbStore(config, 1, callback, null);
    try {
        ValuesBucket values = new ValuesBucket();
        //向学生信息表中插入 3 条记录
        values.putInteger("stuId", 2112);
        values.putString("stuName", "Mike");
        values.putString("stuClass", "计算机 1 班");
        store.insert("studentInfo", values);
        //要记得清空,否则会出现错误
        values.clear();
        values.putInteger("stuId", 2113);
        values.putString("stuName", "Jay");
        values.putString("stuClass", "软工 1 班");
        values.clear();
        values.putInteger("stuId", 2115);
        values.putString("stuName", "Mike");
        values.putString("stuClass", "计算机 2 班");
        store.insert("studentInfo", values);
        values.clear();
        values.putInteger("stuId", 2117);
        values.putString("stuName", "Lee");
        values.putString("stuClass", "计算机 4 班");
        store.insert("studentInfo", values);
        //定义要查询的字段
        String[] columns = new String[]{"stuId", "stuName", "stuClass"};
        //搜索姓名是 Mike 的学生按照学号升序输出
        RdbPredicates rdbPredicates = new RdbPredicates("studentInfo")
                .equalTo("stuName","Mike").orderByAsc("stuId");
        ResultSet resultSet = store.query(rdbPredicates, columns);
```

```
                //对查询结果进行迭代,并输出查询结果中的所有数据
                while(resultSet.goToNextRow()) {
                    HiLogs.info("学生学号:"+resultSet.getInt(0)+",学生姓名:"
                        + resultSet.getString(1)+ ",学生所在班级:"
                        + resultSet.getString(2));
                }
                store.close();//关闭数据库
            } catch (Exception e) {
                HiLogs.info("sqliteWithPredicates:error:" + e.getMessage());
            }
        }
    }
```

运行程序,并单击"使用谓词操作 SQLite 数据库"按钮,会在 HiLog 视图中输出如图 8-4 所示的查询结果。可以看到查询结果输出了两个叫 Mike 的学生信息,并按照学号大小正序输出。

图 8-4 使用谓词查询数据表的结果

8.2.3 使用事务

事务是用户定义的一个数据库操作序列,这些操作要么全做,要么全不做,是一个不可分割的工作单位。如果只执行一条 insert 语句,是可以保证数据一致性的。一旦 SQL 语句执行出错,数据库就会进行回滚,即取消所有的数据修改。也就是说,单条 SQL 语句不存在数据一致性问题。但如果连续执行多条 SQL 语句,该操作就失去了原子性。例如,先执行 insert 语句,然后再次执行一条 insert 语句。由于某些原因,执行第二条 insert 语句时抛出异常,但要求是两条语句要么都要执行成功,要么都失败。在这种情况下,第一条 insert 语句执行成功,第二条 insert 语句执行失败,必须要对这个数据库进行回滚。

为了解决数据库的事务问题,HarmonyOS 提供事务机制,来保证用户多条操作的正确性。对单条数据进行数据库操作时,一般无须开启事务,而插入的数据量比较大时,开启事务可以保证数据的准确性。如果中途操作出现失败,会自动执行回滚操作。通过调用 RdbStore.beginTransaction()方法开启事务,然后使用 RdbStore.endTransaction()方法结束事务;如果事务执行成功,那么所有改动就会保存到数据库;如果事务出错,那么夹在 RdbStore.beginTransaction()方法和 RdbStore.endTransaction()方法之间的所有修改数据库的操作都会回滚。HarmonyOS 的事务默认是回滚的,所以并没有提供回滚方法,而是提供了一个 Rdbstore.markAsCommit()方法。当所有的数据库操作执行成功后,通过该方法

标记事务成功,这时数据会被提交到数据库。如果没有调用该方法,那么事务就会回滚。

HarmonyOS 同时提供 RdbStore.beginTransactionWithObserver() 方法开始事务,该方法可以对事务的执行过程进行监控。该方法需要传入一个事务观察者对象。观察者对象需要实现 TransactionObserver 接口,该接口提供了 onBegin()、onCommit() 和 onRollback() 3 个方法,分别用于对事务开始、事务提交和事务回滚事件进行响应。

【例 8.6】 向学生信息数据库中插入多条记录。因为学生的学号必须是唯一的,所以在执行第二条 insert 语句时数据库会出错。因为对这两条 insert 语句使用了事务,所以第一条学生数据不会被录入数据库中,数据库会进行回滚操作,最后事务观察者对象可以对事务的执行过程进行响应。

```java
public class SQLiteTransactionDemo {
    //事务观察者
    private static class StudentTransactionObserver implements TransactionObserver {
        @Override
        public void onBegin() {
            HiLogs.info("开始事务");
        }
        @Override
        public void onCommit() {
            HiLogs.info("事务提交");
        }
        @Override
        public void onRollback() {
            HiLogs.info("开始回滚");
        }
    }
    public static void useTransaction(Context context) {
        RdbStore store;
        StoreConfig config = StoreConfig
            .newDefaultConfig("transaction.sqlite");
        RdbOpenCallback callback = new RdbOpenCallback() {
            @Override
            public void onCreate(RdbStore store) {
                store.executeSql("CREATE TABLE IF NOT EXISTS studentInfo
                    (stuId INTEGER PRIMARY KEY, stuName VARCHAR(20),
                    stuClass VARCHAR(20))");
                Tools.print("学生信息表创建成功");
            }
            @Override
            public void onUpgrade(RdbStore store, int oldVersion,
                int newVersion) {}
        };
        DatabaseHelper helper = new DatabaseHelper(context);
        store = helper.getRdbStore(config, 1, callback, null);
        try {
            String insertSQL = "insert into studentInfo(stuId,
```

```
            stuName, stuClass) values(?,?,?)";
        //开启一个事务,并通过前面自定义的观察者监听事务
        store.beginTransactionWithObserver(new
        TestTransactionObserver());
        store.executeSql(insertSQL, new Object[]{2112, "Jay",
        "计算机1班"});
        //这条语句会抛出异常,因为学号出现了重复的情况
        store.executeSql(insertSQL, new Object[]{2113, "Lee", "软工一班"});
        //如果成功执行所有的SQL语句,在这边做成功提交的标记
        //否则成功了也会回滚
        store.markAsCommit();
    } catch (Exception e) {
        HiLogs.info("事务出错了需要回滚:error:" + e.getMessage());
    } finally {
        store.endTransaction();
        store.close();
    }
    }
}
```

运行程序,然后单击"录入学生数据"按钮,在 HiLog 视图中会输出如图 8-5 所示的信息。

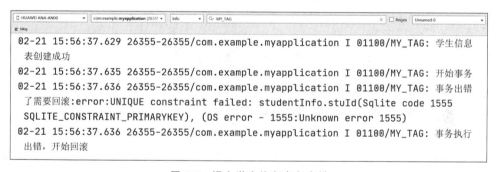

图 8-5 提交学生信息事务出错

从控制台输出信息可以看到这个事务出错了,未能成功提交到数据库。查看输出结果可以看到 HarmonyOS 输出了具体的出错原因:UNIQUE constraint failed,即违反了字段的唯一性,出错的字段是 studentInfo 表的 stuId 字段。修改完错误之后,再次执行代码,执行结果如图 8-6 所示。

图 8-6 提交学生信息事务成功

从输出结果可以看到事务执行成功,对数据库进行查询,学生数据已经全部存到数据库中。

8.3 对象关系映射数据库

对象关系映射(Object Relational Mapping,ORM)数据库是一款基于 SQLite 的数据库框架,屏蔽了底层 SQLite 数据库的 SQL 操作,针对实体和关系提供了增、删、改、查等一系列的面向对象接口。应用开发者不必再去编写复杂的 SQL 语句,以操作对象的形式来操作数据库,有助于提升效率,让使用者更加专注于业务开发。对象关系映射数据库有 3 个主要部分,分别是数据库、实体对象以及对象数据操作接口。数据库被开发者用@Database 注解,且继承了 OrmDatabase 类,对应关系型数据库。实体对象被开发者用@Entity 注解,且继承了 OrmObject 类,对应关系型数据库中的表。对象数据操作接口包括数据库操作的入口 OrmContext 类和谓词接口(OrmPredicate)。

相比于直接使用接口开发,使用对象关系映射数据库需要修改配置,即要修改 build.gradle 文件。build.gradle 文件位于项目的 entry 中,具体有以下 3 种情况:

(1) 如果使用注解处理器的模块为"com.huawei.ohos.hap"模块,则需要在模块的"build.gradle"文件的 ohos 节点中添加以下配置:

```
compileOptions{
    annotationEnabled true
}
```

(2) 如果使用注解处理器的模块为"com.huawei.ohos.library"模块,则需要在模块的"build.gradle"文件的 dependencies 节点中配置注解处理器。

在 HUAWEI SDK 中的 Sdk/java/x.x.x.xx/build-tools/lib/目录中有 3 个 jar 包,分别为"orm_annotations_java.jar""orm_annotations_processor_java.jar"和"javapoet_java.jar"。将目录的这 3 个 jar 包导进项目,复制到＜HarmonyOS 工程根目录＞/entry/libs 目录下,jar 包位置如图 8-7 所示,并且在 build.gradle 中添加配置:

```
dependencies {
    compile files("orm_annotations_java.jar 的路径",
    "orm_annotations_processor_java.jar 的路径",
    "javapoet_java.jar 的路径")
    annotationProcessor files("orm_annotations_java.jar 的路径",
    "orm_annotations_processor_java.jar 的路径","javapoet_java.jar 的路径")
}
```

注意,一定要将能使用注解配置打开,也就是在 ohos 节点中添加以下配置:

```
compileOptions{
    annotationEnabled true
}
```

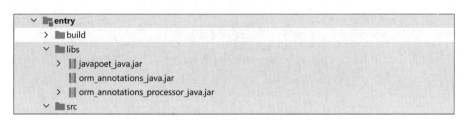

图 8-7 jar 包放置的位置截图

（3）如果使用注解处理器的模块为"java-library"模块，则需要在模块的"build.gradle"文件的 dependencies 节点中配置注解处理器，并导入"ohos.jar"。

```
dependencies {
    compile files("ohos.jar 的路径",
    "orm_annotations_java.jar 的路径",
    "orm_annotations_processor_java.jar 的路径",
    "javapoet_java.jar 的路径")
    annotationProcessor files("orm_annotations_java.jar 的路径",
    "orm_annotations_processor_java.jar 的路径",
    "javapoet_java.jar 的路径")
}
```

做完这些准备工作后，就可以使用注解来修改实体类了。目前支持的注解如表 8-3 所示。

表 8-3 鸿蒙 Orm 支持的注解

注 解 名 称	注 解 描 述
@Database	被@Database 注解且继承了 OrmDatabase 的类对应数据库类
@Entity	被@Entity 注解且继承了 OrmObject 的类对应数据表类
@Column	被@Column 注解的变量对应数据表的字段
@PrimaryKey	被@PrimaryKey 注解的变量对应数据表的主键
@ForeignKey	被@ForeignKey 注解的变量对应数据表的外键
@Index	被@Index 注解的内容对应数据表索引的属性

【例 8.7】 使用 ORM 创建学生信息数据库 student.sqlite 和学生信息表 studentInfo，并添加多条学生信息，修改其中 1 条记录中的学生名字，然后输出结果。本例使用的是第二种引入注解的方式。

首先，创建 studentInfo 表的实体类。

```
@Entity(tableName = "studentInfo")
public class StudentInfo extends OrmObject{
    @PrimaryKey(autoGenerate = true)
    private int stuId;
    private String stuName;
```

```java
    private String stuClass;
    //getter()和setter()方法可以用工具生成
}
```

然后，创建 student.db 对应的实体类。

```java
//指定数据库中有一个 StudentInfo 表
@Database(entities = {StudentInfo.class}, version=1)
    public abstract class Student extends OrmDatabase {
}
```

最后，使用前面创建的两个实体类以及 ORM 相关的 API 操作 SQLite 数据库。

```java
public class OrmTestDemo {
    public static void testOrm(Context context) {
        DatabaseHelper helper = new DatabaseHelper(context);
        //打开学生数据库
        OrmContext ormContext = helper.getOrmContext("Student", "student.db",
            Student.class);
        try {
            //插入多条记录,两个学生名字相同
            StudentInfo studentInfo1 = new StudentInfo(2112, "Jack",
                "计算机 1 班");
            ormContext.insert(studentInfo1);
            StudentInfo studentInfo2 = new StudentInfo(2113, "Mike",
                "计算机 1 班");
            ormContext.insert(studentInfo2);
            StudentInfo studentInfo3 = new StudentInfo(2114, "Mike",
                "计算机 3 班");
            ormContext.insert(studentInfo3);
            //调用 flush()方法会将内存中的数据一次性写入数据库
            ormContext.flush();
            //查看数据
            OrmPredicates predicates = ormContext.where(StudentInfo.class);
            predicates.equalTo("stuName", "Mike");
            List<StudentInfo> studentInfos = ormContext.query(predicates);
            for(int i = 0; i < studentInfos.size(); i++) {
                HiLogs.info("学生学号: " + studentInfos.get(i).getStuId()
                +",学生姓名: "+studentInfos.get(i).getStuName()
                +",学生班级: "+studentInfos.get(i).getStuClass());
            }
            //修改学号为 2113 的学生姓名
            StudentInfo studentInfo = studentInfos.get(0);
            studentInfo.setStuName("Tommy");
            ormContext.update(studentInfo);
            ormContext.flush();
            OrmPredicates predicatesAfter = ormContext
                .where(StudentInfo.class);
```

第8章 数据管理

```
        predicatesAfter.equalTo("stuId", 2113);
        List<StudentInfo> studentInfos2 = ormContext
            .query(predicatesAfter);
        //查看数据是否发生改变
        HiLogs.info("修改之后的数据");
        for(int i = 0; i < studentInfos2.size(); i++) {
            HiLogs.info("学生学号: " + studentInfos.get(i).getStuId()
            +",学生姓名: "+studentInfos.get(i).getStuName()
            +",学生所在班级: "+studentInfos.get(i).getStuClass());
        }
    } catch (Exception e) {
        HiLogs.info("Orm出错" + e.getMessage());
    } finally {
        ormContext.close();//关闭数据库
    }
}
```

运行程序，单击"使用对象关系映射（ORM）"按钮，会生成一个 student.db 文件。然后查看 HiLog 控制台，会看到控制台输出如图 8-8 所示的数据。可以看到插入两条姓名是 Mike 的学生记录，然后通过 update() 函数修改其姓名并将修改后的结果输出。

图 8-8　使用 Orm 插入数据并且修改学生姓名结果

8.4　分布式文件

　　分布式是 HarmonyOS 区别于别的操作系统的特色之一，其中最容易使用的功能就是分布式文件。分布式文件是指依赖于分布式文件系统，分散存储在多个用户设备上的文件，应用间的分布式文件目录互相隔离，不同应用的文件不能互相访问。分布式文件服务能够为用户设备中的应用程序提供多设备之间的文件共享能力，支持相同账号下同一应用文件的跨设备访问，应用程序可以不感知文件所在的存储设备，就能够在多个设备之间无缝获取文件。

　　分布式文件的同步工作完全由操作系统负责，应用只要用于存储分布式文件的目录，然后将文件写入该目录，HarmonyOS 就会自动将该文件同步到其他 HarmonyOS 设备上。

同一个应用在多个设备上的分布式目录是相同的。

分布式文件的使用有以下要求：

（1）程序如需使用分布式文件服务完整功能，需要添加后面提到的权限。

（2）多个设备需要登录相同华为账号，然后打开多个设备的蓝牙，或将多个设备接入同一无线局域网（WLAN），才能实现文件的分布式共享。

（3）当多台设备对同一文件并发写操作时有数据冲突，后写会覆盖先写，应用需要主动保证时序控制并发流程。

（4）应用访问分布式文件时，如果文件所在设备离线，文件不能访问。网络情况差时，访问存储在远端的分布式文件时，可能会长时间不返回或返回失败，开发应用时需要考虑这种场景的处理。

（5）当两台设备有同名文件时，同步元数据时会产生冲突。冲突解决策略：①本地跟远端冲突，远端文件被重命名，看到的同名文件是本地同名文件，远端文件被重命名；②远端多个设备冲突，以接入本设备 ID 为顺序，显示设备 ID 小的同名文件，其他文件被依次重命名；③如果组网场景，目录树下已经有远端文件，创建同名文件，提示文件已存在；④冲突文件显示_conflict_dev 后依次加 id；⑤同名目录之间仅融合不存在冲突，文件和远端目录同名冲突，远端目录后缀加_remote_directory。

分布式文件的关键是使用 Context.getDisributedDir()方法获取分布式目录，但在读写该目录中的文件之前，先要在 config.json 文件中添加如下权限：

```
"reqPermissions":
    {"name": "ohos.permission.DISTRTBUTED_DATASYNC"},
],
```

然后还需要用 Java 代码对分布式特性进行授权，所以需要编写下面的授权方法。

```java
private void requestPermission() {
    String[] permission = {"ohos.permission.DISTRIBUTED_DATASYNC"};
    List<String> applyPermissions = new ArrayList<>();
    for(String element: permission) {
        if(verifySelfPermission(element) != 0) {
            if(canRequestPermission(element)) {
                applyPermissions.add(element);
            } else {}
        } else {}
    }
    requestPermissionsFromUser(applyPermissions.toArray(new String[0]), 0);
}
```

该方法写在 Ability 中即可，在 Ability 的 onstart()方法的开始部分调用 requestPermission()方法进行授权。

【例 8.8】 将 hello.txt 文件写入一部 HarmonyOS 手机的分布式文件存储目录，然后用另一部 HarmonyOS 手机读取 hello.txt 文件的内容，并将该文件的内容输出到 HiLog 视图中。

```
public class DistributedTetsFile {
    //将数据写入分布式文件
    public static void writeFile(Context context) {
        //获取分布式文件存储目录
        File distDir = context.getDistributedDir();
        String filePath = distDir + File.separator + "hello.txt";
        FileWriter fileWriter = null;
        try {
            fileWriter = new FileWriter(filePath, false);
            fileWriter.write("Hello World");
        } catch (IOException e) {
            HiLogs.info("写入分布式文件出错");
        }
    }
}
//读取分布式文件中的数据
public static void readFile(Context context) {
    File distDir = context.getDistributedDir();
    String filePath = distDir + File.separator + "hello.txt";
    char[] buffer = new char[11];
    try {
        FileReader fileReader = new FileReader(filePath);
        fileReader.read(buffer);
        HiLogs.info("设备 B 输出结果:" + String.valueOf(buffer));
        fileReader.close();
    } catch (IOException e) {
        e.printStackTrace();
    }
}
```

运行本例至少需要两部 HarmonyOS 手机（或其他 HarmonyOS 设备），假设这两部 HarmonyOS 手机是 A 和 B。首先在手机 A 上运行程序，然后单击"写入数据"按钮，接下来在手机 B 上运行程序再单击"读出数据"按钮，会看到 HiLog 视图中输出如图 8-9 和图 8-10 所示的信息。

图 8-9　设备 A 输出分布式文件的内容

可以发现在手机 A 和手机 B 的分布式文件存储目录下，都有一个名为 hello.txt 的文件，该文件是 HarmonyOS 自动同步的。显然可以发现两个设备的分布式文件路径是相

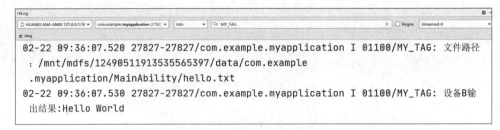

图 8-10　设备 B 输出分布式文件的内容

同的。

8.5　分布式数据

分布式数据与分布式文件类似，也可以在 HarmonyOS 设备之间同步数据。分布式数据服务（Distributed Data Service，DDS）为应用程序提供不同设备间数据库数据分布式的能力。通过调用分布式数据接口，应用程序将数据保存到分布式数据库中。通过结合账号、应用和数据库三元组，分布式数据服务对属于不同应用的数据进行隔离，保证不同应用之间的数据不能通过分布式数据服务互相访问。在通过可信认证的设备间，分布式数据服务支持应用数据相互同步，为用户提供在多种终端设备上最终一致的数据访问体验。

8.5.1　同步数据

分布式数据可以在两部或多部设备之间同步 key-value 格式的数据。key-value 格式数据适合不涉及过多数据关系和业务关系的业务数据存储，比 SQL 数据库存储拥有更好的读写性能，同时因其在分布式场景中降低了解决数据库版本兼容问题的复杂度，和数据同步过程中冲突解决的复杂度而被广泛使用。分布式数据库也是基于 key-value 格式数据，对外提供 key-value 格式的访问接口。

下面是使用分布式数据在多个设备之间同步数据的步骤。

（1）在 config.json 中添加 permission 权限，要和添加在 abilities 同一目录层级。

```
"reqPermissions":[
    {"name": "ohos.permission.DISTRTBUTED_DATASYNC"},
],
```

（2）应用启动时，需要弹出授权提示框，请求用户进行授权。根据配置构造分布式数据库管理类实例，即 KvManager 对象。

```
KvManagerConfig myConfig = new KvManagerConfig(context);
KvManage kvManager = Factory.getInstance().createKvManager(myConfig);
```

（3）获取/创建单版本分布式数据库，即 SingleKvStore 对象。

```
Options myOptions = new Options();
myOptions.setCreateIfMissing(true).setEncrypt(false)
    .setKvStoreType(KvStoreType.SINGLE_VERSION);
```

```
String storeId = "data";
SingleKvStore singleKvStore = kvManager.getKvStore(myOptions, storeId);
```

(4)订阅分布式数据变化。首先客户端需要实现 KvStoreObserver 接口,然后构造并注册 KvStoreObserver 实例。

```
class KvStoreObserverClient implements KvStoreObserver {
    @Override
    public void onChange(ChangeNotification notification) {
        HiLogs.info("数据发生变化。");
    }
}
//以下是注册 KvStoreObserver 的代码
KvStoreObserver kvStoreObserverClient = new KvStoreObserverClient();
singleKvStore.subscribe(SubscribeType.SUBSCRIBE_TYPE_ALL,
kvStoreObserverClient);
```

(5)在发送数据方,需要使用下面的代码写入数据。

```
String dataKey = "Job";
String dataValue = "worker";
singlekvStore.putString(dataKey, dataValue);
```

(6)在发送数据方,当写入数据后,需要使用下面的代码根据设备 ID 同步分布式数据。

```
List<DeviceInfo> deviceInfos = kvManager.
getConnectedDevicesInfo(DeviceFilterStrategy.NO_FILTER);
List<String> deviceIds= new ArrayList<>();
for(DeviceInfo deviceInfo: deviceInfoList) {
    deviceIdList.add(deviceInfo.getId());
}
singleKvStore.sync(deviceIds, SyncMode.PUSH_ONLY);
```

【例 8.9】 使用分布式数据在 A 和 B 两个设备上同步数据。在 A 设备上同步数据,B 设备就会接收到这些数据,并输出到 HiLog 视图中。在 B 设备上同步数据,A 设备也会接收到这些数据,并输出到 HiLog 视图中。

```
//该类同时用于数据发送端和数据接收端
public class KvStoreObserverClient implements KvStoreObserver {
    private SingleKvStore singleKvStore;
    private KvManager kvManager;
    public KvStoreObserverClient (Context context) {
        KvManagerConfig kvManagerConfig = new KvManagerConfig(context);
        kvManager = KvManagerFactory.getInstance()
                .createKvManager(kvManagerConfig);
        Options CREATE=new Options();
        CREATE.setCreateIfMissing(true).setEncrypt(false)
```

```java
            .setKvStoreType(KvStoreType.SINGLE_VERSION);
        String storeID = "testDb";
        singleKvStore = kvManager.getKvStore(CREATE, storeID);
        singleKvStore.subscribe(SubscribeType.SUBSCRIBE_TYPE_ALL, this);
    }
    public void onChange(ChangeNotification changeNotification) {
        //读取这些数据并且输出到 HiLog 视图当中
        HiLogs.info("onChange:" + singleKvStore.getString("location"));
    }
    //获取全部在线设备
    public List<DeviceInfo> getAvailableDeviceIds() {
        List<DeviceInfo> deviceInfoList = kvManager.
                getConnectedDevicesInfo(DeviceFilterStrategy.NO_FILTER);
        if(deviceInfoList.isEmpty()) {
            return new ArrayList<>();
        }
        return deviceInfoList;
    }
    //写入数据,并将这些数据同步到所有可用的设备
    public void writeData() {
        String key = "location";
        String value = "北京";
        //写入数据
        singleKvStore.putString(key, value);
    }
    public void readData() {
        //读出 key-value
        String key = singleKvStore.getString("location");
        HiLogs.info(key);
    }
}
```

运行本例至少需要两台 HarmonyOS 设备,假设是 A 设备和 B 设备。在 A 设备上单击"写入数据"按钮,在 B 设备单机读出数据,在 HiLog 会看到如图 8-11 和图 8-12 所示的信息。在 B 设备上单击"写入数据"按钮,在 A 设备的 HiLog 视图中也会输出同样的信息。在使用 DevEco Studio 时可以使用 HiLog 最顶部三角箭头切换设备的 HiLog 窗口。

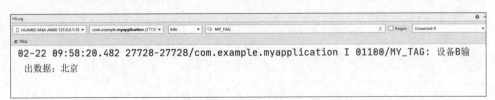

图 8-11　B 设备输出数据

注意,分布式数据也同样需要加 8.4 节的权限,否则无法同步分布式数据。

图 8-12　监听器对写入动作进行响应

8.5.2　用谓词查询分布式数据

HarmonyOS 还提供了一组谓词 API(类似于 ORM 的谓词 API)，用于查询分布式数据。但要注意，目前查询的都是 key-value 形式的数据。现在将数据集从 HarmonyOS A 同步到 HarmonyOS B，并在 B 设备中使用谓词搜索特定的记录，步骤如下：

（1）创建 Schema。单版本数据库支持在创建和打开数据库时指定 Schema，数据库根据 Schema 定义感知 KV 记录的 value 格式，以实现对 value 值结构的检查，并基于 value 中的字段实现索引建立和谓词查询。Schema 相当于数据库中的表，所以为数据集创建一个表(Schema)的代码如下：

```
Schema mySchema = new Schema();
mySchema.setSchemaMode(SchemaMode.COMPATIBLE);
```

（2）创建记录集的字段，并添加到 Schema 中。记录集的字段通常包含字段名、字段类型以及其他属性，如是否为空，默认值等。一个 FieldNode 对象表示一个字段，创建字段的基本方式如下：

```
FieldNode stuNameNode = new FieldNode("stuName");
stuNameNode.setType(FieldValueType.STRING);
//字段不许为空
idNode.setNullable(false);
//为 Schema 添加字段
mySchema.getRootFieldNode().appendChild(stuNameNode);
```

（3）创建索引列表，并且添加到 Schema 中。为了让查询更有效率，通常会在经常查询的字段上创建索引。一个记录集可以有 1 个或多个索引，由一个 List<String>对象表示，代码如下：

```
List<String> indexes = new ArrayList<>();
indexes.add("$.location");
myschema.setIndexes(indexes);
```

（4）设置 Schema 选项。通过 Options 对象可以对 Schema 进行配置，通常的配置如下：

```
Options myOptions = new Options();
//1.检查数据库不存在时是否创建
//2.设置自动同步
```

```
//3.设置数据库是否加密
myOptions.setCreateIfMissing(true).setAutosync(false).setEncrypt(false)
.setkvstoreType
(KvStoreType.SINGLE_VERSION);
options.setSchema(schema);
```

(5)创建数据库(Singlekvstore 对象)。这里说的数据库相当于内存数据库,一个数据库需要用一个字符串作为唯一标识,如"world"。

```
KvManagerConfig myConfig = new KvManagerConfig(this);
KvManger myKvManager =KvManagerFactory.getInstance()
.createkvManager(myConfig);
Singlekvstore mySinglekvstore =myKvManager.getkvStore(options, "world");
```

这里的 Singlekvstore 对象就表示一个数据库,world 相当于数据库名。数据库名字不能重复,否则会抛出异常。

(6)创建观察者类,并且订阅。如果数据需要从 A 设备同步到 B 设备,那么在 B 设备中需要使用观察者类来截获同步过来的数据,观察者类必须实现 KvStoreObserver 接口,实现代码如下:

```
public class DistributedDataobserver implements KvStoreObserver{
    //当 B 设备接收到同步过来的数据时,onChange()方法会被调用
    @Override
    public void onChange(ChangeNotification changeNotification){
    //处理业务的代码
    }
}
//订阅代码
mySingleKvStore.subscribe(SubscribeType.SUBSCRIBE_TYPE_ALL,
new DistributedDataobserver());
```

(7)写入数据。不管是普通的数据,还是数据集类型的数据,都可以使用 putXxx()方式写入数据(Xxx 表示 String、Boolean 等)。如果要写入数据集,则要将数据转换成 JSON 格式,然后通过 putString()方法写入数据,代码如下:

```
mySingleKvStore.putString("key_L","{\"location\":\"北京\"}");
```

这里的 key_L 是任意指定的,但是不要和其他 key 重复。后面通过 Object 形式的 JSON 描述一条记录。因为第(1)步只创建了 1 个字段,而且是 String 类型,所以 JSON 对象只包含一个名为 location 的属性。JSON 对象中的属性名、属性个数和类型必须与第(1)步创建的字段名、字段个数和类型相匹配,否则会抛出异常。

(8)同步数据。通过 Singlekvstore.sync()方法可以将数据同步到其他 HarmonyOS 上,同步需要获取其他 HarmonyOS 的 ID,代码如下:

```
//获取所有可用 HarmonyOS 设备的信息
List<DeviceInfo> deviceInfoList = getAvailableDeviceids();
```

```
List<String> deviceIdList = new ArrayList<>();
for(DeviceInfo deviceInfo: deviceInfoList){
    //将设备 ID 添加到 deviceIdList 列表中
    deviceIdList.add(deviceInfo.getDeviceId());
}
//向 deviceIdList 列表中包含的所有设备同步数据
Singlekvstore.sync(deviceIdList, SyncMode.PUSH_ONLY);
```

为了获取各个设备的信息,除了"ohos.permission.DISTRIBUTED_DATASYNC"权限之外,还需要加入以下权限:

```
{"name":"ohos.permission.DISTRIBUTED_DEVICE_STATE_CHANGE"},
{"name":"ohos.permission.GET_DISTRIBUTED_DEVICE_INFO"},
{"name":"ohos.permission.GET_BUNDLE_INFO"}
```

这些权限是专门为了获取设备信息准备的,不需要获取的时候可以不用添加。

(9) 在 B 设备用谓词查询数据。B 设备通过观察者对象的 onChange()方法接收到数据后,可以通过下面的代码查询数据。

```
Query query = Query.select();
//equalTo 表示相等,还有别的谓词可以在官方文档中查询
query.equalTo("$.location", "北京");
//使用谓词查询,entries 包含了所有符合条件的结果
List<Entry> entries = singleKvStore.getEntries(query);
for(Entry entry: entries) {
    //在 HiLog 视图中输出查询到的结果(JSON 字符串)
    HiLog.print("onChange:entries:" + entry.getValue().getstring());
}
```

【例 8.10】 完整地演示用谓词查询分布式数据的全过程。按照以上的步骤,添加 3 个字段(stuId、stuName 和 stuClass),即学号,学生姓名和学生所在班级,并且在 B 设备搜索姓名是"Mike"的学生的记录,并且将输出结果搜索结果输出到 HiLog 视图中。

```
public class StuDistributedData extends Ability {
    private SingleKvStore singleKvStore;
    private KvManager kvManager;
    //获取全部可用设备信息
    public List<DeviceInfo> getAvailableDeviceIds() {
        List<DeviceInfo> deviceInfoList =
            DeviceManager.getDeviceList(DeviceInfo.FLAG_GET_ONLINE_DEVICE);
        if(deviceInfoList.isEmpty()) {
            return new ArrayList<>();
        }
        return deviceInfoList;
    }
    //观察者内部类
```

```java
private class StuDistributedObserver implements KvStoreObserver {
    @Override
    public void onChange(ChangeNotification changeNotification) {
        //因为 onChange()方法在非 UI 线程中运行,所以显示 ToastDialog 信息框必须
        //让其在 UI 线程中运行,因此要使用 getuITaskDispatcher()方法来运行代码
        getUITaskDispatcher().asyncDispatch(new Runnable() {
            @Override
            public void run() {
                //构造谓词查询
                Query query = Query.select();
                //stuname 字段值等于 Mike,这里指定字段,需要前面加"$"
                query.equalTo("$.stuName", "Jack");
                KvStoreResultSet resultset = singleKvStore
                                                .getResultSet(query);
                //在 HiLog 视图中输出所有的查询结果,并且转化成 Object 对象
                while(resultset.goToNextRow()) {
                    String json = resultset.getEntry().getValue()
                                                .getString();
                    //将对象字符串转换为 JSON 对象
                    ZSONObject object = ZSONObject.stringToZSON(json);
                    //获取学生姓名
                    String name = object.getString("stuName");
                    //获取学生班级
                    String stuClass = object.getString("stuClass");
                    HiLogs.info("数据发生变化" + "学生姓名:" + name
                                    + "学生所在班级:" + stuClass);
                }
            }
        });
    }
}
//主要负责创建数据库、表和字段
public void createDataBase() {
    KvManagerConfig config = new KvManagerConfig(this);
    kvManager = KvManagerFactory.getInstance().createKvManager(config);
    //下面的代码创建了 3 个字段
    FieldNode idNode = new FieldNode("stuId");
    idNode.setType(FieldValueType.INTEGER);
    idNode.setNullable(false);
    FieldNode nameNode = new FieldNode("stuName");
    nameNode.setType(FieldValueType.STRING);
    idNode.setNullable(false);
    FieldNode classNode = new FieldNode("stuClass");
    classNode.setType(FieldValueType.STRING);
    //创建表
    Schema schema = new Schema();
    schema.setSchemaMode(SchemaMode.COMPATIBLE);
    List<String> indexes = new ArrayList<>();
```

```
        indexes.add("$.stuName");
        //给表添加索引,索引是学生姓名
        schema.setIndexes(indexes);
        //添加字段
        schema.getRootFieldNode().appendChild(idNode);
        schema.getRootFieldNode().appendChild(nameNode);
        schema.getRootFieldNode().appendChild(classNode);
        //设置 Schema
        Options options = new Options();
        options.setCreateIfMissing(true).setAutoSync(false)
            .setEncrypt(false).setKvStoreType(KvStoreType.SINGLE_VERSION);
        options.setSchema(schema);
        //创建数据库
        String dbName = "Student";
        singleKvStore = kvManager.getKvStore(options, dbName);
        //订阅观察者对象
        singleKvStore.subscribe(SubscribeType.SUBSCRIBE_TYPE_ALL,
                        new StuDistributedObserver());
}
@Override
public void onStart(Intent intent) {
    super.onStart(intent);
    super.setUIContent(ResourceTable.Layout_ability_main);
    //完成初始化工作
    createDataBase();
    Button writeButton = (Button) findComponentById
                    (ResourceTable.Id_button_write);
    if(writeButton != null) {
        writeButton.setClickedListener(new Component.ClickedListener() {
            //写入数据
            @Override
            public void onClick(Component component) {
                try {
                    singleKvStore.putString("key_2112","{\"stuId\":2112,
                        \"stuName\":\"Mike\",\"stuClass\":\"计算机1班\"}");
                    singleKvStore.putString("key_2113","{\"stuId\":2113,
                        \"stuName\":\"Jack\",\"stuClass\":\"计算机2班\"}");
                    singleKvStore.putString("key_2114","{\"stuId\":2114,
                        \"stuName\":\"Mike\", \"stuClass\":\"计算机3班\"}");
                }catch (Exception e) {
                    HiLogs.info("Error:" +e.getMessage());}
            }
        });
    }
    //同步按钮,如果有多台设备可以使用该方法同步到多台设备
    Button syncButton = (Button) findComponentById
    (ResourceTable.Id_syncButton);
    if(syncButton != null) {
```

```java
            syncButton.setClickedListener(new Component.ClickedListener() {
                @Override
                public void onClick(Component component) {
                    try {
                        List<DeviceInfo> deviceInfoList = getAvailableDeviceIds();
                        List<String> deviceIdList = new ArrayList<>();
                        for(DeviceInfo deviceInfo: deviceInfoList) {
                            deviceIdList.add(deviceInfo.getDeviceId());
                        }
                        //同步数据
                        singleKvStore.sync(deviceIdList, SyncMode.PUSH_ONLY);
                        HiLogs.info("数据同步成功!");
                    } catch (Exception e) {
                        HiLogs.info("数据同步失败" + e.getMessage());
                    }
                }
            });
        }
        Button readButton = (Button) findComponentById
            (ResourceTable.Id_readButton);
        if(readButton != null) {
            readButton.setClickedListener(new Component.ClickedListener() {
                //从本地读数据
                @Override
                public void onClick(Component component) {
                    try {
                        Query query = Query.select();
                        query.equalTo("$.stuName","Mike");
                        List<Entry> entries = singleKvStore.getEntries(query);
                        if(entries.isEmpty())
                        {
                            HiLogs.info("未找到学生信息");
                            for(int i = 0; i < entries.size(); i++)
                                String json = entries.get(i).getValue().getString();
                                //将对象字符串转换为 JSON 对象
                                ZSONObject object = ZSONObject.stringToZSON(json);
                                //获取学号
                                String id = object.getString("stuId");
                                //获取学生姓名
                                String name = object.getString("stuName");
                                //获取学生班级
                                String stuClass = object.getString("stuClass");
                                HiLogs.info("找到的学生信息: "+"学号: "+ id + ",学生姓名:"
                                    + name + ",学生所在班级:" + stuClass);}
                    } catch (Exception e) {
                        HiLogs.info("找数据出错"+ e.getMessage());}}
            });
        }
    }
}
```

阅读这段代码，需要了解以下几点：

（1）分布式数据的记录使用 JSON 格式表示，如果要想获取具体字段值，需要使用 JSON 库解析 JSON 字符串，在编写完 JSON 格式数据可以去专门的校验网站进行校验。

（2）因为观察者对象中的 onChange() 方法在非 UI 线程中运行，所以如果想在该方法中访问 UI 件，或显示 ToastDialog 信息框，需要使用 getUITaskDispatcher().asyncDispatch() 方法运行，否则会抛出异常。

（3）在为 Schema 添加索引或使用谓词查询时，指定字段需要在字段名前面加"$"前缀，表示 JSON 中隐含的根节点，这点前面已经强调了。

（4）查询方法有两个：getEntries() 和 getResultSet()，前者以列表形式返回查询结果，后者以 kvstoreresultSet 形式返回查询结果。一般建议使用后者，因为提供更多筛选的方法。

在 A 设备上单机"写入数据"按钮和"同步数据"按钮，并且在 B 设备上单机"读取数据"按钮，会看到如图 8-13～图 8-15 所示的结果。

图 8-13　在 A 设备单击"写入数据"结果，查询三次姓名为 Jack 的学生信息

图 8-14　在 A 设备单击"同步数据"结果

图 8-15　在 B 设备单击"读取数据"结果

在同步结束之后可以调用 kvManager 的 closeKvStore() 方法和 deleteKvStore() 方法来关闭和删除数据库。

第 9 章 Data Ability

本章就如何使用 Data Ability 做了说明，Data Ability 是 Ability 的一种，可以支持 HarmonyOS 设备访问本地的数据，也可以通过指定 deviceId 访问对应设备的数据。

通过阅读本章，读者可以掌握：
➢ Data Ability 的概念。
➢ Data Ability 中的 URI 结构。
➢ 如何创建 Data Ability。
➢ 如何通过 Data Ability 访问数据库。
➢ 如何通过 Data Ability 访问文件。
➢ 如何通过 Data Ability 跨设备进行访问。

9.1 Data Ability 概述

Data Ability 的作用主要与 App 对数据的处理有关，使用 Data Ability 的目的就是将数据抽象化。Data Ability 可以支持本地数据的访问，也可以支持跨设备数据的访问，同时提供共享自身数据的方法。同样的 Data Ability 支持两种数据共享的形式：本地数据共享和跨设备数据共享。

Data Ability 可以支持对数据进行增、删、改、查和对文件中的数据进行操作，并提供对应 API。数据的来源可以有很多种，SQLite 数据库、文件、网络等，具体的数据来源由开发者依照实际需求提供。使用 Data Ability 有以下优势：

（1）低耦合。由于 Data Ability 支持多种数据源，所以在实际应用场景中数据源的切换会比较灵活方便。开发者只需要关注 Data Ability 内部的具体逻辑实现即可，无需关注输入输出的结构和参数，实现模块解耦。

（2）易维护。实际应用场景中的数据来源可能会有多个，涉及到的数据处理逻辑可能会非常复杂。复杂的逻辑会带来复杂的代码，往往会导致代码的维护变得异常艰难。使用 Data Ability 可以对这些复杂的代码进行封装，在使用中只需调用 Data Ability 中封装好的方法，隔离了代码的复杂性，使代码的维护变得更容易。

9.2 Data Ability 中的 URI

要使用 Data Ability 去访问数据，不管是访问本地数据还是跨设备访问数据，都需要用到 URI(Uniform Resource Identifier)来标识需要访问的数据，格式如下：

```
Scheme://[authority]/<path>[?query][#fragment]
```

用"< >"括起来的部分是必须的,用"[ ]"括起来的部分是可选的,其中各部分描述如下:

(1) Scheme:协议方案名,这一部分固定为 dataability,表示访问的是 Data Ability。

(2) authority:设备 ID,跨设备访问场景下,代表目标设备的 ID。本地设备场景下,则不需要填写。

(3) path:资源的路径信息,代表特定资源的位置信息。

(4) query:用于查询参数。

(5) fragment:用于指定要访问的子资源。

9.3 创建 Data Ability

开发者创建 Data Ability 有两种方式,可以手动创建,也可以使用 DevEco Studio 进行自动创建,推荐使用 DevEco Studio 自动进行 Data Ability 的创建。

9.3.1 手动创建 Data Ability

Data Ability 是 Ability 的一种,所以 Data Ability 在创建时必须继承 Ability 类。创建 Data Ability 最简单的方式就是创建一个继承 Ability 类的空类。

创建一个简单的 Data Ability,类名为 StudentDataAbility,代码如下:

```
public class StudentDataAbility extends Ability {
}
```

创建了 StudentDataAbility 以后需要在 config.json 文件中对 StudentDataAbility 进行注册,config.json 中的注册信息如下:

```
{
    "permissions": [
        "com.example.myApplication.data.DataAbilityShellProvider.PROVIDER"
    ],
    "name": "com.example.myApplication.data.StudentDataAbility",
    "type": "data",
    "uri":  "dataability://com.example.myApplication.data
.StudentDataAbility"
}
```

其中的参数信息说明如下:

(1) name:注册的 Data Ability 名,需要给出完整的引用路径。

(2) type:Ability 的类型,这里设置为 data。

(3) uri:对外提供的访问路径,全局唯一。

(4) permissions:访问该 Data Ability 时需要申请的访问权限。

9.3.2 自动创建 Data Ability

DevEco Studio 可以支持开发者进行 Data Ability 的自动创建。使用 DevEco Studio 进行 Data Ability 的自动创建更加快捷且不易出错，推荐使用自动创建的方式来生成 Data Ability。

开发者要使用 DevEco Studio 创建一个 Data Ability，首先选中需要创建 Data Ability 的包，右击鼠标，在菜单中依次选择"New"→"Ability"→"Empty Data Ability"菜单项，如图 9-1 所示。

单击"Empty Data Ability"菜单后，会弹出如图 9-2 所示的"New Ability"窗口。在窗口中填入需要创建的 Data Ability 类名和包名，单击"Finish"按钮完成创建，DevEco Studio 即会在对应包下生成 Data Ability。

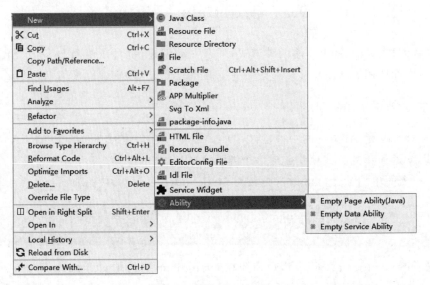

图 9-1 创建 Data Ability

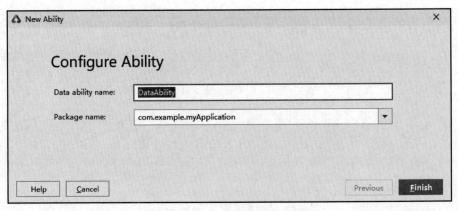

图 9-2 New Ability 窗口

DevEco Studio 自动创建 Data Ability 的同时也会在 config.json 配置文件中对创建的 Data Ability 进行注册，开发者无需手动再去进行注册。

9.3.3 创建 DataAbilityHelper

DataAbilityHelper 类支持对本地提供的共享数据或跨设备提供的共享数据进行访问，可以作为客户端和提供数据一方的 Data Ability 进行通信，提供方的 Data Ability 接收到请求后，依据请求进行数据处理并返回处理结果。

DataAbilityHelper 类的实例化在 Data Ability 的调用方进行。提供数据的待访问 Data Ability 声明了访问需要权限的情况下，访问此 Data Ability 需要在配置文件中对需要的权限进行声明。

```
"reqPermissions": [
    {
        "name": "com.example.myApplication.DataAbility.DATA"
    },
    {
        "name": "ohos.permission.READ_USER_STORAGE"
    },
    {
        "name": "ohos.permission.WRITE_USER_STORAGE"
    }
]
```

DataAbilityHelper 可以使用 creator()方法来进行实例化。

```
DataAbilityHelper helper = DataAbilityHelper.creator(this);
```

9.4 Data Ability 访问数据库

要使用 Data Ability 访问数据库，大致需要经历初始化数据库连接、在 Data Ability 中实现数据库操作方法、使用 DataAbilityHelper 类对 Data Ability 进行调用 3 个步骤。

(1) 建立数据库连接。

对数据库进行访问的第一步便是建立数据库连接。HarmonyOS 会在应用启动时调用 Data Ability 中的 onStart()方法来对 Data Ability 进行初始化。在 Data Ability 初始化过程中，可以同时进行数据库连接的创建，以便进行后续的数据库操作。如何创建数据库连接已经在第 8 章中详细说明，这里不再赘述。

(2) 在 Data Ability 中实现数据库操作方法。

完成数据库连接后就需要为 Data Ability 实现数据库操作方法。Data Ability 提供了 query()、insert()、delete()、update()接口，覆盖了数据库操作的增、删、改、查，开发者在实际应用场景中使用时只需要针对需求使用 ORM 操作数据库自行实现对应功能即可。

(3) 使用 DataAbilityHelper 类对 Data Ability 进行调用。

DataAbilityHelper 类可以作为调用方对 Data Ability 进行调用。对应 Data Ability 提供的增、删、改、查接口，DataAbilityHelper 类同样也为开发者提供了对应的数据库操作方法，详细内容如表 9-1 所示。

表 9-1　DataAbilityHelper 类提供的数据库操作方法

方　　法	描　　述
ResultSet query(Uri uri,String[] columns,DataAbilityPredicates predicates)	查询数据库
int insert(Uri uri,ValuesBucket value)	向数据库中插入单条数据
int batchInsert(Uri uri,ValuesBucket[] values)	向数据库中插入多条数据
int delete(Uri uri,DataAbilityPredicates predicates)	删除一条或多条数据
int update(Uri uri,ValuesBucket value,DataAbilityPredicates predicates)	更新数据库
DataAbilityResult[] executeBatch(ArrayList<DataAbilityOperation> operations)	批量操作数据库

以 insert()方法为例，DataAbilityHelper 执行 insert()方法需要在参数中指定调用的 Data Abiltiy 的 URI 和需要插入数据库的具体数据内容。执行 insert()方法后，DataAbilityHelper 会和 URI 指定的 Data Ability 进行通信，告知 Data Ability 需要做的数据库操作和对应的具体数据内容，Data Ability 完成以后返回操作结果给 DataAbilityHelper。

【例 9.1】 构建一个名为 StudentDataAbility 的 Data Ability，要求 StudentDataAbility 创建名为 student.sqlite 的数据库和 student 表（用于存储学生信息）。StudentDataAbility 实现对 student 表的增、删、改、查方法，然后通过 DataAbilityHelper 对 StudentDataAbility 进行调用。

首先在 StudentDataAbility 中的 onStart()方法里实现数据库的连接和初始化，然后在增、删、改、查方法中实现对应的数据库操作，代码如下：

```java
public class StudentDataAbility extends Ability {
    private static final HiLogLabel LABEL_LOG = new HiLogLabel(3, 0xD001100,
                                                                "MY_TAG");
    private static final String DATABASE_NAME = "Student.db";
    private static final String DATABASE_NAME_ALIAS = "Student";
    private static OrmContext ormContext = null;

    public void initDB(Context context) { //初始化数据库
        StoreConfig storeConfig = StoreConfig.newDefaultConfig
        ("student.sqlite");
        RdbOpenCallback rdbOpenCallback = new RdbOpenCallback() {
            @Override
            public void onCreate(RdbStore rdbStore) {
                rdbStore.executeSql("CREATE TABLE IF NOT EXISTS student (id
                    INTEGER PRIMARY KEY, name VARCHAR(20), gender VARCHAR(20),
                    major VARCHAR(20), classNo VARCHAR(20), msg TEXT)");
            }
            @Override
            public void onUpgrade(RdbStore rdbStore, int i, int i1) {
            }
```

```java
        };
        DatabaseHelper helper = new DatabaseHelper(context);
        RdbStore store = helper.getRdbStore(storeConfig, 1, rdbOpenCallback,
            null);
        try {
            RdbPredicates rdbPredicates = new RdbPredicates("student");
            store.delete(rdbPredicates);
            store.close();
        } catch (Exception e) {
        }
        ormContext = helper.getOrmContext(DATABASE_NAME_ALIAS,
                DATABASE_NAME, StudentDB.class);
    }

    @Override
    public void onStart(Intent intent) {
        super.onStart(intent);
        initDB(this);
    }

    @Override            //查询方法实现
    public ResultSet query(Uri uri, String[] columns, DataAbilityPredicates
    predicates) {
        OrmPredicates ormPredicates = DataAbilityUtils.createOrmPredicates
        (predicates, Student.class);
        return ormContext.query(ormPredicates, columns);
    }

    @Override            //插入方法实现
    public int insert(Uri uri, ValuesBucket value) {
        Student student = new Student();
        student.setId(value.getInteger("id"));
        student.setName(value.getString("name"));
        student.setGender(value.getString("gender"));
        student.setMajor(value.getString("major"));
        student.setClassNo(value.getString("classNo"));
        student.setMsg(value.getString("msg"));
        try {
            ormContext.insert(student);
            ormContext.flush();
        } catch (Exception e) {
            return 0;
        }
        return 1;
    }

    @Override            //删除方法实现
    public int delete(Uri uri, DataAbilityPredicates predicates) {
```

```
        OrmPredicates ormPredicates =DataAbilityUtils.createOrmPredicates
        (predicates, Student.class);
        int value = ormContext.delete(ormPredicates);
        ormContext.flush();
        DataAbilityHelper.creator(this, uri).notifyChange(uri);
        return value;
    }

    @Override          //更新方法实现
    public int update(Uri uri, ValuesBucket value, DataAbilityPredicates
    predicates) {
        OrmPredicates ormPredicates =DataAbilityUtils.createOrmPredicates
        (predicates, Student.class);
        int result = ormContext.update(ormPredicates, value);
        ormContext.flush();
        DataAbilityHelper.creator(this, uri).notifyChange(uri);
        return result;
    }
}
```

StudentDataAbility 中涉及的数据库类 StudentDB 如下：

```
@Database(entities = {Student.class}, version = 1)
public class StudentDB extends OrmDatabase {
    @Override
    public int getVersion() {
        return 0;
    }

    @Override
    public RdbOpenCallback getHelper() {
        return null;
    }
}
```

StudentDataAbility 中涉及的实体类 Student 如下：

```
@Entity(tableName = "student")
public class Student extends OrmObject {
    @PrimaryKey(autoGenerate = true)
    private Integer id;
    private String name;
    private String gender;
    private String major;
    private String classNo;
    private String msg;
    //省略 get()和 set()方法
}
```

使用 DataAbilityHelper 类对 StudentDataAbility 进行调用，对数据库进行增、删、改、查操作，代码如下：

```java
public class MainAbilitySlice extends AbilitySlice {
    private static final HiLogLabel LABEL_LOG = new HiLogLabel(3, 0xD001100,
                                    "MY_TAG");

    @Override
    public void onStart(Intent intent) {
        super.onStart(intent);
        super.setUIContent(ResourceTable.Layout_ability_main);
        Button button = (Button) findComponentById
                    (ResourceTable.Id_operationButton);
        if(button != null) {
            button.setClickedListener(new Component.ClickedListener() {
                @Override
                public void onClick(Component component) {
                    try {
                        operationDataAbility ();
                    } catch (Exception e) {
                        e.printStackTrace();
                    }
                }
            });
        }
    }

    public void operationDataAbility() throws DataAbilityRemoteException {
        Uri uri = Uri.parse("dataability:///com.example.myApplication
                    .StudentDataAbility");
        DataAbilityHelper helper = DataAbilityHelper.creator
                            (MainAbilitySlice.super.getContext());
        //插入数据库操作
        ValuesBucket studentOne = new ValuesBucket();
        studentOne.putInteger("id", 1);
        studentOne.putString("name", "小明");
        studentOne.putString("gender", "男");
        studentOne.putString("major", "计算机");
        studentOne.putString("classNo", "1班");
        studentOne.putString("msg", "三好学生");
        helper.insert(uri, studentOne);

        ValuesBucket studentTwo = new ValuesBucket();
        studentTwo.putInteger("id", 2);
        studentTwo.putString("name", "小红");
        studentTwo.putString("gender", "女");
        studentTwo.putString("major", "计算机");
```

```java
        studentTwo.putString("classNo", "1班");
        studentTwo.putString("msg", "三好学生");
        helper.insert(uri, studentTwo);
        //更新数据库操作
        DataAbilityPredicates updatePredicates = new DataAbilityPredicates();
        updatePredicates.equalTo("id", 2);
        ValuesBucket updateStudent = new ValuesBucket();
        updateStudent.putString("msg","普通学生");
        helper.update(uri, updateStudent, updatePredicates);
        //删除数据库操作
        DataAbilityPredicates deletePredicates = new DataAbilityPredicates();
        deletePredicates.equalTo("id", 1);
        helper.delete(uri, deletePredicates);
        //查询数据库操作
        DataAbilityPredicates queryPredicates = new DataAbilityPredicates();
        queryPredicates.greaterThan("id", 0);
        String[] columns = new String[]{"id", "name", "gender", "major",
                        "classNo", "msg"};
        ResultSet resultSet = helper.query(uri, columns, queryPredicates);
        if(resultSet != null) {
            while(resultSet.goToNextRow()) {
                HiLog.info(LABEL_LOG, String.format("id=%s, name=%s,
                    gender=%s, major=%s, classNo=%s, msg=%s",
                    resultSet.getInt(0),resultSet.getString(1),
                    resultSet.getString(2), resultSet.getString(3),
                    resultSet.getString(4), resultSet.getString(5)));
            }
            resultSet.close();
        }
    }
}
```

运行结果如图 9-3 所示。

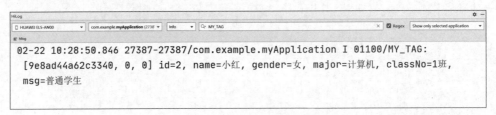

图 9-3 Data Ability 访问数据库结果

9.5 Data Ability 访问文件

Data Ability 不仅可以访问数据库，也可以访问文件。Data Ability 中提供了 openFile() 接口，允许开发者根据实际需求自己实现文件操作。和访问数据库一样，DataAbilityHelper

类中也有对应的 openFile（Uri uri，String mode）方法可供开发者使用，openFile()方法中的参数描述如下：

（1）uri：确定目标资源路径。

（2）mode：指定文件的打开方式，包含"r"（读）、"w"（写）、"rw"（读写）、"wt"（覆盖写）、"wa"（追加写）和"rwt"（覆盖写且可读）6种方式。

访问文件示例：

```
FileDescriptor fd = helper.openFile(uri, "r");
FileInputStream fis = new FileInputStream(fd);
```

【例9.2】 创建一个名为 FileDataAbility 的 Data Ability，在 FileDataAbility 初始化时生成一个文件 student.txt，用于记录学生信息。在 FileDataAbility 中实现文件的读取，然后使用 DataAbilityHelper 类对 FileDataAbility 进行调用，输出 student.txt 文件中记录的学生信息。

首先在 FileDataAbility 的 onStart()方法中创建 student.txt 文件并初始化学生信息，在 openFile()方法中实现文件访问操作，代码如下：

```
public class FileDataAbility extends Ability {
    private static final HiLogLabel LABEL_LOG = new HiLogLabel(3, 0xD001100,
                                                            "MY_TAG");

    @Override
    public void onStart(Intent intent) {
        super.onStart(intent);
        createFile();
    }

    private void createFile() {
        String filename = "student.txt";
        File dataDirFile = getDataDir();
        try {
            FileWriter fw = new FileWriter(dataDirFile.getAbsoluteFile()
                                    + "/" + filename);
            fw.write("id=1\r\nname=小明\r\ngender=男\r\nmajor=计算机\r\n
                    classNo=1 班\r\nmsg=三好学生\r\n"+"id=2\r\n
                    name=小红\r\ngender=女\r\nmajor=计算机\r\n
                    classNo=2 班\r\nmsg=普通学生");
            fw.close();
        } catch (Exception e) {
        }
    }

    @Override
    public FileDescriptor openFile(Uri uri, String mode) throws
    FileNotFoundException {
        File dataDirFile = getDataDir();
        MessageParcel messageParcel = MessageParcel.obtain();
```

```java
            File file =new File(dataDirFile.getAbsolutePath()
                    +"/"+uri.getDecodedPathList().get(1));
            if(mode== null || "r".equals(mode)){
                file.setReadOnly();
            }
            try {
                FileInputStream fis = new FileInputStream (file);
                FileDescriptor fd = fis.getFD();
                return messageParcel.dupFileDescriptor(fd);
            } catch (Exception e) {
            }
            return null;
        }
}
```

调用端使用 DataAbilityHelper 类对 FileDataAbility 进行调用,读取 student.txt 文件中的内容并输出,代码如下:

```java
public class MainAbilitySlice extends AbilitySlice {
    private static final HiLogLabel LABEL_LOG = new HiLogLabel(3, 0xD001100,
"MY_TAG");

    @Override
    public void onStart(Intent intent) {
        super.onStart(intent);
        super.setUIContent(ResourceTable.Layout_ability_main);
        Button button = (Button) findComponentById
                (ResourceTable.Id_operationButton);
        if(button != null) {
            button.setClickedListener(new Component.ClickedListener() {
                @Override
                public void onClick(Component component) {
                    try {
                        readFile();
                    } catch (Exception e) {
                        e.printStackTrace();
                    }
                }
            });
        }
    }

    private void readFile() throws Exception {
        Uri uri = Uri.parse("dataability:///com.example.myApplication
                .FileDataAbility/student.txt");
        DataAbilityHelper helper = DataAbilityHelper.creator
                    (MainAbilitySlice.super.getContext());
```

```
            try {
                FileDescriptor fd = helper.openFile(uri, "r");
                if(fd == null) {
                    throw new Exception("文件不存在");
                }
                FileReader fr = new FileReader(fd);
                BufferedReader br = new BufferedReader(fr);
                String line = "";
                String studentInfo = "";
                while((line = br.readLine()) != null) {
                    if(line.contains("msg")){
                        studentInfo += line + "\r\n";
                    }else{
                        studentInfo += line + ", ";
                    }
                }
                br.close();
                HiLog.info(LABEL_LOG, studentInfo);
            }catch (Exception e){
            }
        }
}
```

运行结果如图 9-4 所示。

图 9-4 Data Ability 访问文件结果

9.6 Data Ability 跨设备访问

Data Ability 除了支持对本地的数据进行访问以外，同时还支持数据的跨设备访问。跨设备访问实质上就是对其他设备上的 Data Ability 进行调用，由该设备上的 Data Ability 进行数据操作。

要实现访问其他 HarmonyOS 设备上的 Data Ability，需要在 URI 中指定对应设备的 deviceId，同时还要在待访问设备的 config.json 文件中对注册的配置进行修改。

（1）待访问设备的 Data Ability 中的 visible 设为 true。visible 的默认值为 false，默认情况下跨设备是无法访问 Data Ability 的。

```json
{
    "permissions": [
        "com.example.myApplication.DataAbilityShellProvider.PROVIDER"
    ],
    "name": "com.example.myApplication.StudentDataAbility",
    "icon": "$media:icon",
    "description": "$string:studentdataability_description",
    "type": "data",
    "visible": true,
    "uri": "dataability://com.example.myApplication.StudentDataAbility"
}
```

（2）待访问的 Data Ability 的权限改为系统级。

```json
"defPermissions":[{
    "name": "com.example.myApplication.DataAbilityShellProvider.PROVIDER",
    "grantMode": "system_grant"
}]
```

（3）加入分布式数据同步和设备信息获取权限。

```json
"reqPermissions": [
    {
        "name": "com.example.myApplication.StudentDataAbility.DATA"
    },
    {
        "name": "ohos.permission.GET_DISTRIBUTED_DEVICE_INFO"
    },
    {
        "name": "ohos.permission.ACCESS_DISTRIBUTED_ABILITY_GROUP"
    },
    {
        "name": "ohos.permission.DISTRIBUTED_DATASYNC"
    },
    {
        "name": "ohos.permission.READ_USER_STORAGE"
    },
    {
        "name": "ohos.permission.WRITE_USER_STORAGE"
    },
    {
        "name": "com.example.myApplication.DataAbilityShellProvider.PROVIDER"
    }
]
```

同时在 Data Ability 调用时需要加入动态权限，保证拥有足够的权限去访问设备。

```
private void requirePermission() {
    String[] permission = {
        "ohos.permission.DISTRIBUTED_DATASYNC",
        "ohos.permission.servicebus.ACCESS_SERVICE"
    };
    requestPermissionsFromUser(permission, 0);
}
```

(4) 跨设备访问需要指定 deviceId，使用 DeviceManager.getDeviceList()方法去获取待访问设备的 deviceId，组成一个带 deviceId 的 URI。

```
List<DeviceInfo> deviceList = DeviceManager.getDeviceList(DeviceInfo
                                .FLAG_GET_ONLINE_DEVICE);
String deviceId = deviceList.get(0).getDeviceId();    //获取在线设备的 deviceId
Uri uri = Uri.parse ("dataability://" + deviceId + "/com.example.myApplication
                .StudentDataAbility");                //生成带 deviceId 的 URI
```

【例 9.3】 在 9.4 节中已经实现了 StudentDataAbility，能够对学生信息进行访问，在此基础上对调用方进行修改，实现数据库的跨设备访问。本例需要在两个 HarmonyOS 设备中进行部署。

StudentDataAbility 的实现可以参照 9.4 节，这里不再赘述。调用方的代码如下：

```
public class MainAbilitySlice extends AbilitySlice {
    private static final HiLogLabel LABEL_LOG = new HiLogLabel(3, 0xD001100,
                                                    "MY_TAG");

    @Override
    public void onStart(Intent intent) {
        super.onStart(intent);
        super.setUIContent(ResourceTable.Layout_ability_main);
        requirePermission(); //保证跨设备访问的权限
        Button button = (Button) findComponentById
                    (ResourceTable.Id_operationButton);
        if(button != null) {
            button.setClickedListener(new Component.ClickedListener() {
                @Override
                public void onClick(Component component) {
                    try {
                        operationRemoteDatabase();
                    } catch (Exception e) {
                        e.printStackTrace();
                    }
                }
            });
        }
    }
```

```java
    private void requirePermission() {
        String[] permission = {
            "ohos.permission.DISTRIBUTED_DATASYNC",
                "ohos.permission.servicebus.ACCESS_SERVICE"
        };
        requestPermissionsFromUser(permission, 0);
    }

    private void operationRemoteDatabase(){
        List<DeviceInfo> deviceList = DeviceManager.getDeviceList(DeviceInfo
                            .FLAG_GET_ONLINE_DEVICE);
        //获取在线设备的 deviceId
        String deviceId = deviceList.get(0).getDeviceId();
        //生成带 deviceId 的 URI
        Uri uri = Uri.parse("dataability://" + deviceId +
                "/com.example.myApplication.StudentDataAbility");
        DataAbilityHelper helper = DataAbilityHelper.creator
                (MainAbilitySlice.super.getContext());
        try {
            ValuesBucket studentBucket = new ValuesBucket();
            studentBucket.putInteger("id", 1);
            studentBucket.putString("name", "小明");
            studentBucket.putString("gender", "男");
            studentBucket.putString("major", "计算机");
            studentBucket.putString("classNo", "1班");
            studentBucket.putString("msg", "三好学生");
            helper.insert(uri, studentBucket);
            studentBucket.putInteger("id", 2);
            studentBucket.putString("name", "小红");
            studentBucket.putString("gender", "女");
            studentBucket.putString("major", "计算机");
            studentBucket.putString("classNo", "1班");
            studentBucket.putString("msg", "普通学生");
            helper.insert(uri, studentBucket);
            studentBucket.putInteger("id", 3);
            studentBucket.putString("name", "小蔡");
            studentBucket.putString("gender", "女");
            studentBucket.putString("major", "计算机");
            studentBucket.putString("classNo", "2班");
            studentBucket.putString("msg", "三好学生");
            helper.insert(uri, studentBucket);

            DataAbilityPredicates predicates = new DataAbilityPredicates();
            predicates.greaterThan("id",0);
            ResultSet resultSet = helper.query(uri, null, predicates);
            while(resultSet.goToNextRow()) {
                HiLog.info(LABEL_LOG, String.format("id=%s, name=%s,
```

```
                gender=%s, major=%s, classNo=%s, msg=%s",r
                esultSet.getInt(0),resultSet.getString(1),
                resultSet.getString(2),resultSet.getString(3),
                resultSet.getString(4),resultSet.getString(5)));
            }
            resultSet.close();
        } catch (Exception e) {
            HiLog.error(LABEL_LOG, e.toString());
        }
    }
}
```

运行结果如图 9-5 所示。

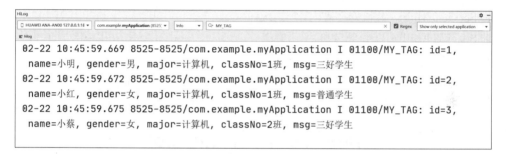

图 9-5 Data Ability 跨设备访问结果

第 10 章 Service Ability

本章就如何使用 Service Ability 做了说明,Service Ability 在 HarmonyOS 中非常重要,主要的作用就是用于后台任务的运行。

通过阅读本章,读者可以掌握:
- Service Ability 的概念。
- Service Ability 的生命周期。
- 如何创建 Service Ability。
- 如何启动和关闭 Service Ability。
- 如何连接 Service Ability。

10.1 Service Ability 概述

Service Ability 属于 Page Ability 中的一种,是非常重要的 Ability,主要作用是用于后台运行任务,本身不提供用户交互界面。Service Ability 可由其他 Ability 启动,Service Ability 启动以后,除非明确关闭或者系统资源紧张自动回收,否则即使用户切换应用,Service Ability 仍旧会继续保持在后台运行。

需要注意的是 Service Ability 是单实例的,一个设备只能允许实例化一个 Service Ability。如果多个 Ability 调用了同一个 Service Ability,那么只有当所有调用的 Ability 全部结束调用以后,Service Ability 才会被系统关闭。

Service Ability 同样也支持本地调用和跨设备调用,允许暴露 API,供其他设备调用。

10.2 Service Ability 的生命周期

Service Ability 本身是 Page Ability 的一种,所以和 Page Ability 一样,Service Ability 也有自己的生命周期。根据调用方法的不同,Service Ability 有两种不同的生命周期。

10.2.1 启动状态的 Service Ability

启动状态的 Service Ability,应用场景是由其他的 Ability 调用 startAbility()方法对 Service Ability 进行创建,实例化以后会一直保持运行。调用方可以使用 stopAbility()来停止 Service Ability,当所有调用方都停止调用以后,系统将会销毁 Service Ability 实例。如图 10-1 所示为启动状态的 Service Ability 的生命周期。

10.2.2 连接状态的 Service Ability

连接状态的 Service Ability,应用场景是由其他的 Ability 作为客户端使用 connectAbility()方法来连接 Service Ability,Service Ability 实例化以后会一直保持运行。客户端可以调用 disconnectAbility() 来断开连接,当所有调用方都停止调用以后,系统将会销毁 Service Ability 实例。如图 10-2 所示为连接 Service Ability 生命周期。

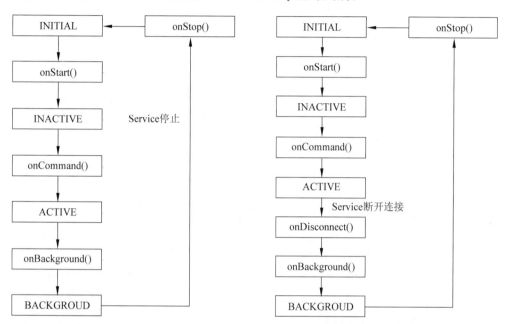

图 10-1　启动状态的 Service Ability 的生命周期　　图 10-2　连接状态的 Service Ability 的生命周期

10.3　Service Ability 的创建

Service Ability 同样也是一种 Ability,这意味着 Service Ability 必须要继承 Ability 类,创建 Service Ability 的同时也必须要实现其相关的生命周期方法,开发者可以根据自身的应用场景来重写这些方法。这些生命周期方法如下:

```
public void onStart(Intent intent);
public void onBackground();
public void onStop();
public void onCommand(Intent intent, boolean restart, int startId);
public IRemoteObject onConnect(Intent intent);
public void onDisconnect(Intent intent);
```

10.3.1　创建 Service Ability

DevEco Studio 提供了自动创建 Service Ability 的功能。首先选中需要创建 Service Ability 的包,IDE 会把 Service Ability 创建到这个包下,单击鼠标右键,在弹出的菜单中选择"New"→"Ability"→"Empty Service Ability"菜单项,如图 10-3 所示。

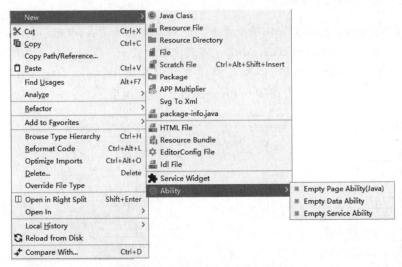

图 10-3　创建 Service Ability 窗口

单击"Empty Service Ability"菜单后,会弹出如图 10-4 所示的 New Ability 窗口。在窗口中填入需要创建的 Service Ability 类名和包名,然后单击"Finish"按钮完成 Service Ability 的创建,IDE 即会在对应包下生成 Service Ability。

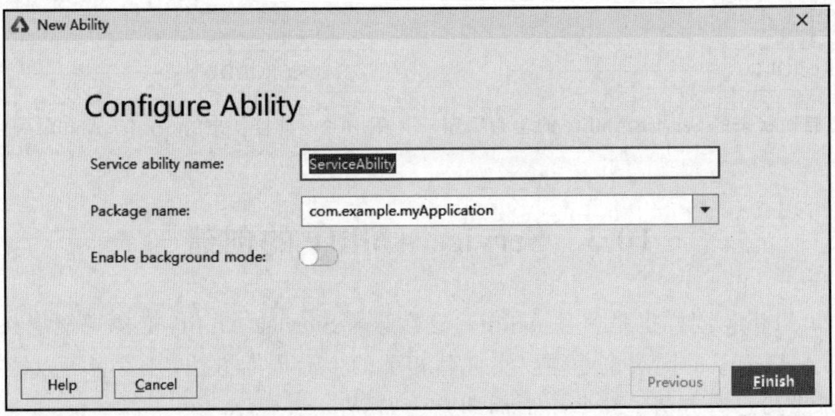

图 10-4　New Ability 窗口

完成创建以后,IDE 会自动生成 Service Ability 并在 config.json 中自动进行注册,注册内容如下:

```
{
    "name": "com.example.myApplication.ServiceAbility",
    "icon": "$media:icon",
    "description": "$string:serviceability_description",
    "type": "service"
}
```

10.3.2 创建前台 Service Ability

在一般的使用场景下，Service Ability 都是采用后台运行的方式运行的，当系统资源紧张的情况下，系统有可能会销毁掉后台运行的 Service Ability 并回收相关资源。在一些特定的场景下，开发者会希望 Service Ability 能一直保持运行而不会被系统销毁回收，为了实现这一点，就需要使用前台 Service Ability。

要让 Service Ability 保持在前台运行，只需要在创建 Service Ability 的时候使用 keepBackgroundRunning() 方法将 Service Ability 与通知进行绑定即可。需要关闭前台 Service Ability 时在 onStop() 方法中调用 cancelBackgroundRunning() 方法即可关闭前台 Service Ability。

使用前台 Service Ability 的代码如下：

```
NotificationRequest request = new NotificationRequest(notificationId);
NotificationRequest.NotificationNormalContent content = new
            NotificationRequest.NotificationNormalContent();
content.setTitle("title").setText("text");

NotificationRequest.NotificationContent notificationContent = new
            NotificationRequest.NotificationContent(content);
request.setContent(notificationContent);

keepBackgroundRunning(notificationId, request); //绑定通知
```

在 config.json 中需要作如下配置：

```
"abilities": [{
    "name": ".ServiceAbility",
    "icon": "$media:icon",
    "description": "$string:serviceability_description",
    "type": "service",
    "visible": true,
    "backgroundModes": ["dataTransfer", "location"]
}],
"reqPermissions": [{
    "name": "ohos.permission.KEEP_BACKGROUND_RUNNING",
    "reason": "get right",
    "usedScene": {
        "ability": [".ServiceAbility"],
        "when": "inuse"
    }
}]
```

10.4　Service Ability 的启动与关闭

10.4.1　启动 Service Ability

在 HarmonyOS 中，如果一个 Ability 需要调用另一个 Ability，可以使用 startAbility() 方法来实现。Service Ability 也是一种 Ability，如果需要启动的是 Service Ability，同样的也可以使用 startAbility() 方法来进行调用。

要使用 startAbility() 方法启动 Service Ability 需要构建一个 Operation 对象来指定待调用的 Service Ability，构造 Operation 对象需要以下几个参数：

（1）DeviceId：表示设备 ID。如果是本地设备，则可以直接留空。如果是远程设备，可通过 ohos.distributedschedule.interwork.DeviceManager 提供的 getDeviceList 获取设备列表。

（2）BundleName：表示包名称。

（3）AbilityName：表示待启动的 Ability 名称。

值得注意的是，在一些资料中启动 Service Ability 使用了 setAction() 方法设定启动的 Service Ability，这样虽然也能够通过编译，但是 setAction() 方法在新版本的 SDK 中已经被注明废弃，不推荐使用。

启动本地 Service Ability 的示例代码如下：

```
Intent intent = new Intent();
Operation operation = new Intent.OperationBuilder().withDeviceId("")
    .withBundleName("com.example.myApplication")
    .withAbilityName("com.example.myApplication.ServiceAbility").build();

intent.setOperation(operation);
startAbility(intent);
```

启动远程设备 Service Ability 的示例代码如下：

```
Intent intent = new Intent();
Operation operation = new
    Intent.OperationBuilder().withDeviceId("deviceId")
    .withBundleName("com.example.myApplication")
    .withAbilityName("com.example.myApplication.ServiceAbility")
    .withFlags(Intent.FLAG_ABILITYSLICE_MULTI_DEVICE).build();

intent.setOperation(operation);
startAbility(intent);
```

10.3 节中提到过 Service Ability 的生命周期方法，第一次调用 Service Ability 时，系统会对 Service Ability 进行实例化，创建实例以后就会调用 onStart() 方法来对 Service Ability 进行初始化，然后回调 onCommand() 方法来启动 Service Ability。如果不是第一次调用，系统则会直接调用 onCommand() 方法来启动 Service Ability。

10.4.2 关闭 Service Ability

关闭 Service Ability 和启动 Service Ability 的方式一样,都要构建一个 Operation 对象来设置待调用的 Service Ability,然后调用 stopAbility()方法停止 Service Ability。

关闭本地 Service Ability 的示例代码如下:

```
Intent intent = new Intent();
Operation operation = new Intent.OperationBuilder().withDeviceId("")
    .withBundleName("com.example.myApplication")
    .withAbilityName("com.example.myApplication.ServiceAbility").build();

intent.setOperation(operation);
stopAbility(intent);
```

关闭远程设备 Service Ability 的示例代码如下:

```
Intent intent = new Intent();
Operation operation = new Intent.OperationBuilder()
    .withDeviceId("deviceId")
    .withBundleName("com.example.myApplication")
    .withAbilityName("com.example.myApplication.ServiceAbility")
.withFlags(Intent.FLAG_ABILITYSLICE_MULTI_DEVICE) .build();

intent.setOperation(operation);
stopAbility(intent);
```

使用 stopAbility()方法停止 Service Ability 以后,系统便会尽快销毁停止的 Service Ability。

【例 10.1】 构建一个 Service Ability 并实现 onStart(),onCommand(),onBackground(),onStop()这 4 个方法,用 startAbility()和 stopAbility()方法启动和停止 Service Ability。

首先在 Service Ability 中实现 onStart(),onCommand(),onBackground(),onStop()方法,为了清晰地反映 Service Ability 所处的生命周期,每个生命周期方法都输出各自所在的生命周期名称,代码如下:

```
public class ServiceAbility extends Ability {
    private static final HiLogLabel LABEL_LOG = new HiLogLabel(3, 0xD001100,
                                                    "MY_TAG");
    @Override
    public void onStart(Intent intent) {
        HiLog.info(LABEL_LOG, "ServiceAbility::onStart"+"---"
            +"执行 onStart()方法");
        super.onStart(intent);
    }

    @Override
    public void onBackground() {
```

```java
        super.onBackground();
        HiLog.info(LABEL_LOG, "ServiceAbility::onBackground"+"---"
                +"执行 onBackground()方法");
    }

    @Override
    public void onStop() {
        super.onStop();
        HiLog.info(LABEL_LOG, "ServiceAbility::onStop"+"---"
                +"执行 onStop()方法");
    }

    @Override
    public void onCommand(Intent intent, boolean restart, int startId) {
        HiLog.info(LABEL_LOG, "ServiceAbility::onCommand"+"---"
                +"执行 onCommand()方法");
    }
}
```

在调用方中单击 startButton 使用 startAbility()方法启动 Service Ability，单击 stopButton 使用 stopAbility()方法停止 Service Ability，代码如下：

```java
public class MainAbilitySlice extends AbilitySlice {
    private static final HiLogLabel LABEL_LOG = new HiLogLabel(3, 0xD001100,
                                                                "MY_TAG");

    @Override
    public void onStart(Intent intent) {
        super.onStart(intent);
        super.setUIContent(ResourceTable.Layout_ability_main);
        Button startButton = (Button) findComponentById
                        (ResourceTable.Id_startButton);
        Button stopButton = (Button) findComponentById
                        (ResourceTable.Id_stopButton);
        Intent myIntent = new Intent();              //创建 Intent

        Operation operation = new Intent.OperationBuilder().withDeviceId("")
            .withBundleName("com.example.myApplication")
            .withAbilityName("com.example.myApplication.ServiceAbility")
            .build();

        myIntent.setOperation(operation);
        if(startButton != null) {
            startButton.setClickedListener(new Component.ClickedListener() {
                @Override
                public void onClick(Component component) {
                    startAbility(myIntent);          //启动服务
```

```
                    HiLog.info(LABEL_LOG, "Service Ability已启动");
                }
            });
        }
        if(stopButton != null) {
            stopButton.setClickedListener(new Component.ClickedListener() {
                @Override
                public void onClick(Component component) {
                    stopAbility(myIntent);          //停止服务
                    HiLog.info(LABEL_LOG, "Service Ability已停止");
                }
            });
        }
    }
}
```

运行代码，单击 startButton，Service Ability 启动，调用 onStart()和 onCommand()方法，输出结果如图 10-5 所示。

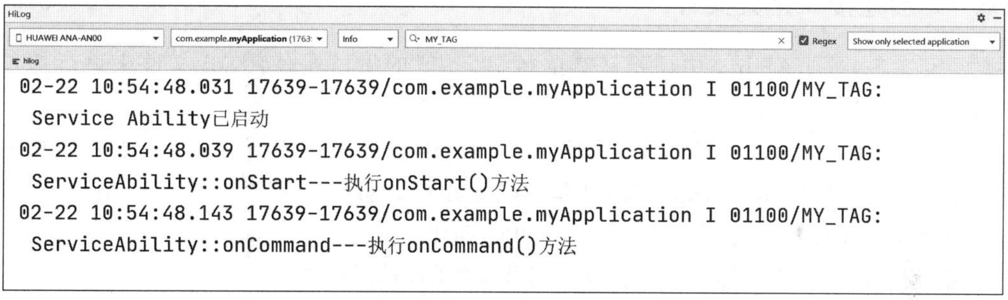
HiLog
HUAWEI ANA-AN00
com.example.myApplication (1763
Info
MY_TAG
Regex
Show only selected application
hilog
02-22 10:54:48.031 17639-17639/com.example.myApplication I 01100/MY_TAG:
Service Ability已启动
02-22 10:54:48.039 17639-17639/com.example.myApplication I 01100/MY_TAG:
ServiceAbility::onStart---执行onStart()方法
02-22 10:54:48.143 17639-17639/com.example.myApplication I 01100/MY_TAG:
ServiceAbility::onCommand---执行onCommand()方法

图 10-5　启动 Service Ability

再次单击 startButton，调用 Service Ability，由于 Service Ability 已经在运行中，直接调用 onCommand()方法，输出结果如图 10-6 所示。

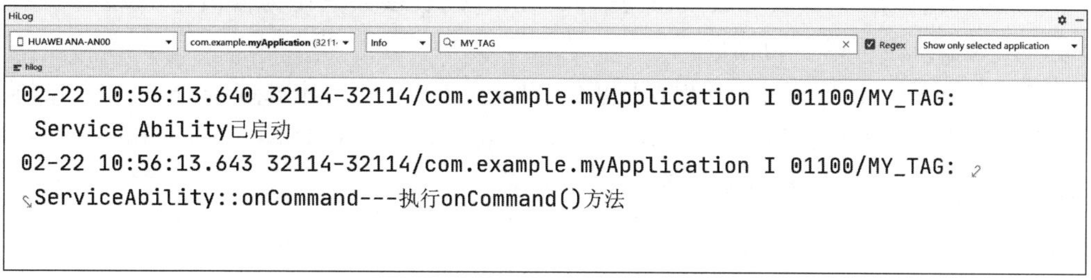
HiLog
HUAWEI ANA-AN00
com.example.myApplication (3211
Info
MY_TAG
Regex
Show only selected application
hilog
02-22 10:56:13.640 32114-32114/com.example.myApplication I 01100/MY_TAG:
Service Ability已启动
02-22 10:56:13.643 32114-32114/com.example.myApplication I 01100/MY_TAG:
ServiceAbility::onCommand---执行onCommand()方法

图 10-6　调用 Service Ability

单击 stopButton，Service Ability 关闭，调用 onBackground()方法和 onStop()方法，输出结果如图 10-7 所示。

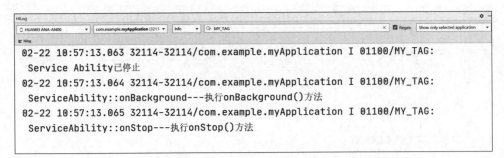

图 10-7　关闭 Service Ability

10.5　Service Ability 的连接

Service Ability 可以向外部暴露一些 API,这些 API 可以供其他的 Ability 调用。本地调用和跨设备调用的实现方法基本是一样的,唯一的区别在于本地调用时不需要去指定 deviceId,跨设备调用时需要指定对应设备的具体 deviceId。

10.5.1　创建接口定义文件

要定义 Service Ability 中需要暴露的接口,一般的做法是使用 idl 文件来进行接口的定义。

idl 文件的位置在 src/main/idl 子目录下。DevEco Studio 支持创建 idl 文件,选中需要创建 idl 文件的目录,单击鼠标右键,在弹出的菜单中选择"New"→"Idl File"菜单项,按照 DevEco Studio 的提示进行即可完成 idl 文件的创建。

在 idl 文件中可以添加要暴露的接口,idl 文件的内容如下:

```
interface com.example.myApplication.StudentLoginInterface {
    String studentLogin([in] String studentName);
}
```

参数类型前的[in]代表参数只作为输入,[out]代表参数只作为输出,[inout]代表参数既可以作为输入也可以作为输出。

编译整个工程,IDE 就会自动根据提供的 idl 文件在 entry/build/generated/source/idl 路径下生成对应接口名的 XXXInterface.java,XXXInterfaceProxy.java,XXXInterfaceStub.java 这 3 个文件(XXX 是 idl 文件命名的接口名)。

10.5.2　连接 Service Ability

当其他的 Ability 需要与 Service Ability 进行交互时,可以使用 connectAbility()方法来进行连接。HarmonyOS 提供了 IAbilityConnection 类来实现 Ability 的连接,IAbilityConnection 类提供了 onAbilityConnectDone()和 onAbilityDisconnectDone()两个方法来处理连接和断开连接的回调。

连接 Service Ability 的代码如下:

```java
private IAbilityConnection connection = new IAbilityConnection() {
    @Override
    public void onAbilityConnectDone(ElementName elementName,
    IRemoteObject iRemoteObject, int resultCode) {
        //连接到 Service 的回调
        //在这里开发者可以拿到服务端传过来 IRemoteObject 对象
        //从中解析出服务端传过来的信息
    }

    @Override
    public void onAbilityDisconnectDone(ElementName elementName,
                                        int resultCode) {
        //断开与连接的回调
    }
};
connectAbility(intent, connection); //连接 Service
```

在其他 Ability 连接 Service Ability 时，Service Ability 需要在执行 onConnect() 方法时返回一个 IRemoteObject 对象，IRemoteObject 对象的作用就是用于定义与 Service Ability 进行通信的接口。开发者可以通过继承 10.5.1 节中用 idl 编译生成的 XXXStub.java 文件来创建对应的 IRemoteObject 并依据实际需求实现需要的方法，代码如下：

```java
private static class StudentLoginRemoteObject extends
StudentLoginInterfaceStub {
    public StudentLoginRemoteObject(String descriptor) {
        super(descriptor);
    }

    @Override
    public String studentLogin(String studentName) throws RemoteException {
        return String.format("你好，%s 欢迎使用学生信息管理系统", studentName);
    }
}
```

需要停止 Service Ability 连接时，可以使用 disconnectAbility() 方法断开 Service Ability 的连接，当 Service Ability 的所有连接都断开以后，系统会停止 Service Ability 并对资源进行回收。

【例 10.2】 构建一个 Service Ability 实现 studentLogin 接口，在 Service 连接时返回学生登录的欢迎信息。在其他的 Ability 中用连接 Service Ability 的方式调用 studentLogin 接口。

在 idl 文件中定义 studentLogin 接口，代码如下：

```
interface com.example.myApplication.StudentLoginInterface {
    String studentLogin([in] String studentName);
}
```

在 Service Ability 中实现 studentLogin 接口，在连接 Service Ability 时返回对应的

IRemoteObject 对象，代码如下：

```java
public class ServiceAbility extends Ability {
    private static final HiLogLabel LABEL_LOG = new HiLogLabel(3, 0xD001100,
                                                "MY_TAG");

    @Override
    public void onStart(Intent intent) {
        super.onStart(intent);
    }

    @Override
    public void onBackground() {
        super.onBackground();
    }

    @Override
    public void onStop() {
        super.onStop();
    }

    @Override
    public void onCommand(Intent intent, boolean restart, int startId) {
    }

    @Override
    public IRemoteObject onConnect(Intent intent) {
        HiLog.info(LABEL_LOG, "ServiceAbility::onConnect");
        return new StudentLoginRemoteObject("studentLogin");
    }

    @Override
    public void onDisconnect(Intent intent) {
        HiLog.info(LABEL_LOG, "ServiceAbility::onDisconnect");
    }

    private static class StudentLoginRemoteObject extends
    StudentLoginInterfaceStub {
        public StudentLoginRemoteObject(String descriptor) {
            super(descriptor);
        }
        @Override
        public String studentLogin(String studentName) throws
                                        RemoteException {
            return String.format("你好,%s 欢迎使用学生信息管理系统",
                            studentName);
        }
    }
}
```

在调用方中单击 startButton 对 Service Ability 进行连接并调用 studentLogin()方法，输出返回结果。单击 stopButton 断开 Service Ability 连接，代码如下：

```java
public class MainAbilitySlice extends AbilitySlice {
    private static final HiLogLabel LABEL_LOG = new HiLogLabel(3, 0xD001100,
                                                "MY_TAG");
    private IAbilityConnection connect;
    @Override
    public void onStart(Intent intent) {
        super.onStart(intent);
        super.setUIContent(ResourceTable.Layout_ability_main);
        Intent myIntent = new Intent();
        Operation operation = new Intent.OperationBuilder().withDeviceId("")
            .withBundleName("com.example.myApplication")
            .withAbilityName("com.example.myApplication.ServiceAbility")
            .build();

        myIntent.setOperation(operation);
        Button startButton = (Button) findComponentById
                (ResourceTable.Id_startButton);
        Button stopButton = (Button) findComponentById
                (ResourceTable.Id_stopButton);
        if(startButton != null) {
            startButton.setClickedListener(new Component.ClickedListener() {
                @Override
                public void onClick(Component component) {
                    startAbility(myIntent);   //启动服务
                    connect = createConnect(myIntent);
                    HiLog.info(LABEL_LOG, "service已启动");
                }
            });
        }
        if(stopButton != null) {
            stopButton.setClickedListener(new Component.ClickedListener() {
                @Override
                public void onClick(Component component) {
                    disconnectAbility(connect);     //停止连接
                    stopAbility(myIntent);          //停止服务
                    HiLog.info(LABEL_LOG, "Service Ability已停止");
                }
            });
        }
    }
    private IAbilityConnection createConnect(Intent intent) {
        IAbilityConnection connection = new IAbilityConnection() {
            private StudentLoginInterfaceProxy studentLoginInterfaceProxy;
            @Override
```

```java
            public void onAbilityConnectDone(ElementName elementName,
                    IRemoteObject iRemoteObject, int resultCode) {
                //连接到 Service 的回调
                studentLoginInterfaceProxy = new StudentLoginInterfaceProxy
                (iRemoteObject);
                try {
                    String loginStr = studentLoginInterfaceProxy.studentLogin
                            ("小明");
                    HiLog.info(LABEL_LOG, "Service Ability已连接");
                    HiLog.info(LABEL_LOG, loginStr);
                } catch (RemoteException e) {
                }
            }

            @Override
            public void onAbilityDisconnectDone(ElementName elementName,
                                        int resultCode) {
                //断开与连接的回调
                studentLoginInterfaceProxy = null;
            }
        };
        connectAbility(intent, connection); //连接 Service
        return connection;
    }
}
```

运行输出结果如图 10-8 所示。

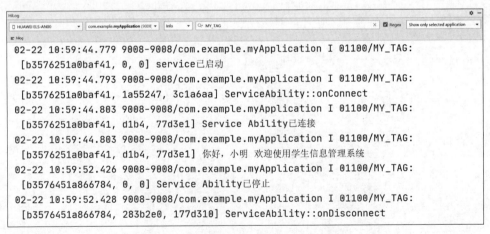

图 10-8　连接 Service Ability 运行结果

第 11 章　工　大　通

本章开始将利用前 10 章所学知识开发一个完整的案例,同时也会介绍一些新的知识,例如,在 HarmonyOS App 中调用第三方开源组件。本章案例是一款管理高校学生通行码的 App,根据学生是否正确提交当日信息判断是否可以持绿码进出校园,综合使用了布局、组件、对话框、数据管理等基础知识,逻辑相对简单,对于初学 HarmonyOS App 开发人员比较友好,能够很好地利用已掌握的知识开发第一个比较完整的案例。

通过阅读本章,读者可以掌握:
➢ 布局与组件的综合使用。
➢ 数据管理。
➢ 表单收集与提交。

11.1　功能需求分析

在开发一款 App 前,明确 App 的需求是必要的,需求分析是软件开发过程中的重要步骤,也是软件生存周期的一个重要环节,该阶段是分析 App 在功能上需要"实现什么",而不是考虑如何去"实现"。本章案例有两个主要功能:每日一报和通行码领取。接下来将分别详细介绍。

11.1.1　每日一报

用户通过"每日一报"功能模块上报当日的健康状况基本信息,包括用户的基本信息以及用户体温和校园通行码颜色等其他信息。用户在此提交的信息需要存储到对象关系映射数据库(ORM 数据库)中,在后续用户领取通行码时作为判断依据。表单涉及了不同的数据类型,因此需要选用合适的组件供用户输入对应的数据类型的信息。

在开发 App 时,用户与 App 的交互体验也应该被考虑到,由于用户是看不到后台和数据库交互的操作的,因此,当用户填写信息并提交到后台进行处理后,需要一个反馈信息给到用户,如"提交成功"等提示信息,避免用户因看不到后台插入数据的操作而重复提交表单信息。

11.1.2 通行码领取

通行码是用户进出校园的凭证，共有两种状态：红码和绿码。红码表示该用户当日未正确填写每日一报，不可进入校园；绿码表示该用户已正确填写每日一报，可以进入校园。通行码领取页面还应该展示用户的部分信息作为标识，如用户姓名、校园通行码颜色等。

11.1.3 数据管理

用户上报的信息需要存储在本地数据库中，同时，在获取通行码时也需要从数据库中取出用户上报的信息，本案例采用 HarmonyOS 为开发者提供的对象关系映射数据库，ORM 数据库基于 SQLite 的数据库框架，但屏蔽了底层的 SQLite 数据库的 SQL 操作，即不需要开发者编写复杂的 SQL 语句，HarmonyOS 为开发者提供了针对实体和关系的增、删、改、查等一系列的面向对象的接口，读者只需调用这些接口即可对数据库进行操作。

11.1.4 业务流程图

在明确 App 的功能需求后，还需要将功能之间的逻辑用流程图的形式表示出来，开发过程中也应该按此逻辑进行开发，保证功能的完整性，本章案例业务流程图如图 11-1 所示。

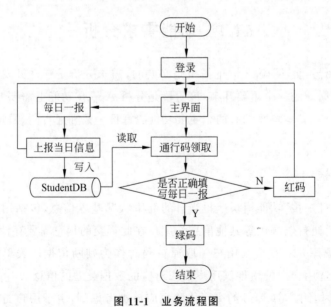

图 11-1 业务流程图

11.1.5 系统构架图

系统架构图描述了整个应用各个功能模块之间的调用关系。本案例系统架构图如图 11-2 所示。

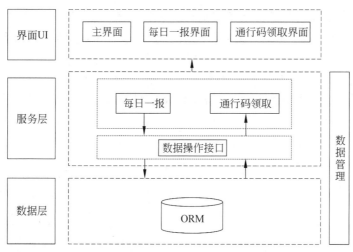

图 11-2 系统架构图

11.2 搭建项目框架

在开发案例之前需要创建项目的基本框架,并配置好 App 的起始页面和 App 图标及名称等信息。

11.2.1 创建项目

打开 DevEco Studio,选择"File"→"New"→"NewProject"菜单项,会显示如图 11-3 所示的"Create Project"对话框。可以看到有多种模板供开发者选用,本章案例只需选择

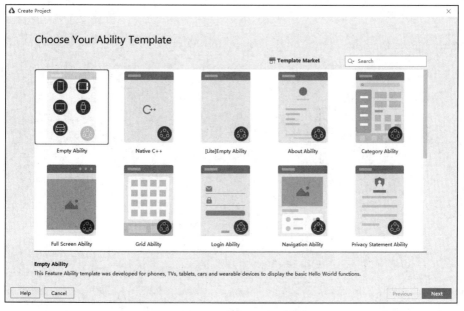

图 11-3 "Create Project"对话框

"Empty Ability"模板即可。选择模板之后配置项目名称、包名等项目信息。注意,开发语言需要选择 Java。这样,一个空的项目已经创建好了,接下来需要对项目进行一些配置。

11.2.2 配置起始页

HarmonyOS 自动创建的项目的起始页默认为 MainAbility,因此需要将项目的起始页修改为本章案例的起始页:登录页面。在 config.json 文件中将 module 属性下的 mainAbility 属性值设置为 LoginAbility,再配置 LoginAbility 的 skills 属性,代码如下:

```
"mainAbility": ".Login.LoginAbility",
"skills": [
    {
        "entities": [
            "entity.system.home"
        ],
        "actions": [
            "action.system.home"
        ]
    }
]
```

11.2.3 配置 App 图标和名称

HarmonyOS 创建项目时会给 App 指定一个默认图标和默认名称,为了和其他应用区分,最好将 App 的图标以及名称更改。首先准备一个图标资源 app_logo.png,然后在 config.json 文件中进行配置,注意,App 图标以及名称的配置需要在起始页的 Page Ability 的配置项中进行配置才能生效,代码如下所示。在进行以下配置后,启动应用可以看到如图 11-4 所示的效果。

```
"abilities": [
    {
        "skills": [
            {
                "entities": [
                    "entity.system.home"
                ],
                "actions": [
                    "action.system.home"
                ]
            }
        ],
        "orientation": "unspecified",
        "name": ".Login.LoginAbility",
        "icon": "$media:app_logo",
        "description":
            "$string:mainability_description",
```

```
            "label":
            "$string:industrial_university_campaign",
            "type": "page",
            "launchType": "standard",
        }
```

图 11-4　配置 App 图标和名称

11.3　界 面 设 计

　　UI 界面是用户与 App 进行数据交互的唯一途径,App 需要用户输入哪些信息,用户需要 App 展示哪些信息,这些都需要在 UI 界面上体现。因此,设计 UI 界面通常是在第一步就要完成的。接下来将详细介绍本案例几个重要的 UI 界面。

11.3.1　登录界面

　　登录界面是整个案例的入口界面,每个用户都有一个属于自己的标识(用户名),App通过这个唯一标识对不同的用户数据进行管理。本案例只做教学使用,并未对用户账号进行管理,只是模拟了用户登录的过程,通常,实际应用都会将用户账号信息存到数据库中,并对用户信息进行管理(增删改查等操作),读者可以在此案例基础上增加对用户信息的管理功能。
　　登录界面的设计相对简单,只用到最常用的方向布局以及 Image、TextField、Button 等基本组件,考虑到界面的美观性,可以编写以 shape 标签作为根标签的 XML 资源文件自定义组件样式,在组件中设置背景为自定义的样式文件即可,样式文件是可以复用的,因此,后续涉及相同组件的样式时将不再赘述,登录界面所涉及的样式文件代码如下。登录界面效果如图 11-5 所示,因布局较为简单,代码不再给出,读者可尝试复现该界面。

```xml
background_btn.xml
<?xml version="1.0" encoding="utf-8"?>
<shape
    xmlns:ohos="https://schemas.huawei.com/res/ohos"
    ohos:shape="rectangle">
    <corners ohos:radius="5fp"></corners>
    <solid ohos:colors="#1890ff"></solid>
</shape>
login_input_layout.xml
<?xml version="1.0" encoding="utf-8"?>
<shape
    xmlns:ohos="https://schemas.huawei.com/res/ohos"
    ohos:shape="rectangle">
    <corners ohos:radius="20vp"></corners>
    <solid ohos:colors="#FFE5E4E4"></solid>
</shape>
```

图 11-5　登录界面

11.3.2　主界面

　　主界面需要展示用户姓名以及用户学号,并提供"每日一报"和"通行码领取"两个功能模块的入口,界面的布局也相对简单,注意,因不同机型的界面尺寸、分辨率的不同,在设计界面的布局时最好采用按权重分配的方式设计布局的长宽,让同样的布局文件在不同手机上显示的效果一致。主界面所涉及的组件样式同样只需设置 shape 文件的以下两个属性即可。

```xml
<corners ohos:radius="20vp"></corners>
<solid ohos:colors="#FFE5E4E4"></solid>
```

整个页面分为三个部分,采用垂直布局方式,第一部分用两个 Text 组件展示用户的基本信息,代码如下:

```xml
<DirectionalLayout
    ohos:height="60vp"
    ohos:width="match_parent"
    ohos:orientation="horizontal">
    <Text
        ohos:height="match_parent"
        ohos:width="0vp"
        ohos:weight="3"
        ohos:text="姓名: "
        ohos:text_size="30fp"
        ohos:text_color="#ffffff"/>
    <Text
        ohos:id="$+id:name_show"
        ohos:height="match_parent"
        ohos:width="0vp"
        ohos:weight="7"
        ohos:text_size="30fp"
        ohos:text_color="#ffffff"/>
</DirectionalLayout>
<DirectionalLayout
    ohos:height="60vp"
    ohos:width="match_parent"
    ohos:orientation="horizontal">
    <Text
        ohos:height="match_parent"
        ohos:width="0vp"
        ohos:weight="3"
        ohos:text="学号: "
        ohos:text_size="30fp"
        ohos:text_color="#ffffff"/>
    <Text
        ohos:id="$+id:id_show"
        ohos:height="match_parent"
        ohos:width="0vp"
        ohos:weight="7"
        ohos:text_size="30fp"
        ohos:text_color="#ffffff"/>
</DirectionalLayout>
```

第二部分仅展示了一张图片,此处不进行介绍。

第三部分是两个 Button,作为"每日一报"和"通行码领取"的入口,代码如下:

```xml
<DirectionalLayout
    ohos:height="0vp"
    ohos:weight="2"
    ohos:width="match_parent"
    ohos:orientation="horizontal">
    <Button
        ohos:id="$+id:dailyReport"
        ohos:height="match_parent"
        ohos:width="0vp"
        ohos:weight="1"
        ohos:right_margin="8vp"
        ohos:background_element="$graphic:red_button"
        ohos:text="每日一报"
        ohos:text_alignment="center"
        ohos:text_size="25fp"
        ></Button>
    <Button
        ohos:id="$+id:getPassCode"
        ohos:height="match_parent"
        ohos:width="0vp"
        ohos:weight="1"
        ohos:background_element="$graphic:blue_button"
        ohos:text="校园通行码"
        ohos:text_alignment="center"
        ohos:text_size="25fp"
        ></Button>
</DirectionalLayout>
```

至此完成了主界面的设计,效果如图 11-6 所示。

图 11-6　主界面

11.3.3 每日一报界面

每日一报界面由标题行和一个滚动视图 ScrollView 组成，ScrollView 中有一个表单，用于接收用户输入的信息，表单中包含了 TextFiled、RadioButton 等基本组件，ScrollView 的使用使得整个表单界面可以上下滑动。注意，在使用 ScrollView 时，必须要指定 ScrollView 的高度，不然会出现无法滑动的情况，本案例将 ScrollView 和标题行的高度按权重分配。部分代码如下：

```
<DependentLayout
    ohos:height="0"
    ohos:weight="0.6"
    ohos:width="match_parent"
    ohos:background_element="#ffffff">
    <Button
        ohos:id="$+id:back_to_main"
        ohos:height="30fp"
        ohos:width="30fp"
        ohos:background_element="$media:back"
        ohos:vertical_center="true"
        ohos:left_margin="10fp"/>
    <Text
        ohos:height="30fp"
        ohos:width="85fp"
        ohos:text="每日一报"
        ohos:text_size="20fp"
        ohos:center_in_parent="true"
        /></DependentLayout>
<ScrollView
    ohos:id="$+id:scrollview"
    ohos:height="0"
    ohos:weight="9.4"
    ohos:width="match_parent"
    ohos:background_element="#F8F8F8"
    ohos:bottom_padding="16vp"
    ohos:layout_alignment="horizontal_center">
<!-- 表单布局代码 -->
<!-- …… -->
</ScrollView>
```

表单中需要注意的是 RadioContainer 和 RadioButton 的使用，RadioContainer 包裹下的 RadioButton 只能有一个被选中，即单选按钮。以用户选择性别为例使用 RadioContainer 和 RadioButton 设置单选按钮的代码如下，其他需要用到单选按钮的地方读者可根据如图 11-7 所示的效果图自行实现。

```
<RadioContainer
    ohos:id="$+id:std_gender"
```

```
        ohos:height="match_content"
        ohos:width="match_parent"
        ohos:orientation="vertical"
        ohos:left_margin="20vp"
        ohos:right_margin="20vp"
        ohos:top_margin="5vp"
        ohos:background_element="$graphic:background_input_layout">
    <RadioButton
        ohos:height="match_content"
        ohos:width="match_content"
        ohos:text="男 Male"
        ohos:text_size="13fp"
        ohos:top_margin="10vp"
        ohos:check_element="$graphic:background_radio_btn">
    </RadioButton>
    <RadioButton
        ohos:height="match_content"
        ohos:width="match_content"
        ohos:text="女 Female"
        ohos:text_size="13fp"
        ohos:bottom_margin="10vp"
        ohos:check_element="$graphic:background_radio_btn">
    </RadioButton>
</RadioContainer>
```

图 11-7　每日一报

由于表单的前两项数据是由主界面传递到每日一报界面的，且不允许用户更改，因此需要在 TextFiled 组件属性设置中添加以下代码设置组件为不可单击状态，使得用户无法输入前两项数据。

```
ohos:clickable="false"
```

11.3.4 通行码领取界面

通行码领取界面主要展示用户的基本信息、通行码以及用户填写的每日一报信息的部分信息，如图 11-8 所示。同样将通行码领取界面分为标题行和主内容视图，标题行的实现参考每日一报界面的标题行代码即可，主视图中子组件之间需要互相重叠，即展示通行码的布局需要堆叠在 Image 组件上，因此需要采用位置布局(PositionLayout)自定义子组件的具体位置。本界面中有一个动态变化的时钟，需要用到 Clock 组件，该组件的具体使用如下：

图 11-8　通行码领取

```
<Clock
    ohos:id="$+id:clock"
    ohos:height="80vp"
    ohos:width="250vp"
    ohos:text_color="red"
    ohos:text_alignment="center"
    ohos:layout_alignment="center"
    ohos:time_zone="GMT"
    ohos:mode_24_hour="MM-dd HH:mm:ss"
    ohos:text_size="28fp"/>
```

通行码的颜色由Java代码动态设置,将在之后的小节进行详细介绍。

关于本章案例所有的静态界面的布局已经全部介绍完毕,接下来将对具体功能的实现做详细介绍。

11.4 功 能 实 现

11.4.1 登录功能

本案例的登录功能仅为模拟登录,没有和数据库进行连接,即没有建立用户表来维护用户账号的信息。读者也可以参考11.4.4节的数据管理部分自行创建一个用户账户表,在登录时进行登录校验。模拟登录只需获取用户输入的姓名与学号,在跳转至主界面时将数据传递到主界面即可,登录功能部分代码如下:

```java
public class LoginAbilitySlice extends AbilitySlice {
    @Override
    public void onStart(Intent intent) {
        super.onStart(intent);
        super.setUIContent(ResourceTable.Layout_login);
        //组件初始化
        Button login = (Button) findComponentById
                (ResourceTable.Id_login_button);
        TextField name_textview=(TextField) findComponentById
                (ResourceTable.Id_name_textField);
        TextField id_textview=(TextField) findComponentById(
                ResourceTable.Id_id_text_field);
        //设置按钮监听事件
        login.setClickedListener(new Component.ClickedListener(){
            @Override
            public void onClick(Component component) {
                Intent intent = new Intent();
                Operation operation = new Intent.OperationBuilder()
                            .withBundleName(getBundleName())
                            .withAbilityName("MainAbility")
                            .build();
                intent.setOperation(operation);
                //携带参数 getText()获取输入框的值
                intent.setParam("stdName",name_textview.getText());
                intent.setParam("stdId",id_textview.getText());
                //启动 MainAbility
                startAbility(intent);
            }
        });
    }
}
```

11.4.2 表单信息收集

表单信息收集即获取用户在输入框输入的数据以及选中的单选按钮映射的值，获取输入框的值只需调用 TextFiled.getText() 方法即可。要获取单选按钮对应的值，需要给 RadioContainer 设置状态改变的响应事件 MarkChangedListener()，当选中某一个单选按钮时，在 onCheckedChanged() 方法中将返回一个 int 类型的索引值，因此需要将索引值转换为实际需要的数据，以选择性别为例，代码如下所示，其他相似的数据获取方法不再赘述。

```
RadioContainer std_gender=
(RadioContainer)findComponentById(ResourceTable.Id_std_gender);
std_gender.setMarkChangedListener((radioContainer, index) -> {
    String value = null;
    //将 0 和 1 转换为"male"和"female"
    switch (index){
        case 0: value = "male"; break;
        case 1: value = "female";}
    student.setGender(value);
});
```

用户在填写表单时，需要动态显示与隐藏部分组件，例如，"是否离杭"这个属性，若选择"是"则需要将用于填写"离杭原因"的组件显示，反之，将其隐藏。组件的显示与隐藏需要用到 setVisibility() 方法，具体使用代码如下：

```
RadioContaine std_in_hangzhou;
TextFiled out_hangzhou_reason;
//组件初始化
out_hangzhou_reason =
    (TextField)findComponentById(ResourceTable.Id_out_hangzhou_reason);
std_in_hangzhou =
    (RadioContainer)findComponentById(ResourceTable.Id_std_in_hangzhou);
std_in_hangzhou.setMarkChangedListener(((radioContainer, index) -> {
    Boolean in_hangzhou = false;
    switch (index){
        case 0: in_hangzhou = true;
            //显示填写离杭原因的组件
            out_hangzhou_reason.setVisibility(Component.VISIBLE);
            break;
        case 2: in_hangzhou = false;
            //隐藏填写离杭原因的组件
            out_hangzhou_reason.setVisibility(Component.HIDE);
    }
    student.setIn_hangzhou(in_hangzhou);
}));
```

在选择表单中的"身体症状（是否发热）"属性时，也需要动态显示选择时间的组件，模仿上述代码即可，此处的时间选择框引入了鸿蒙第三方开源组件，如图 11-9 所示，引入第三方

组件需要在 moudle 级别下的 build.gradle 文件中添加依赖,代码如下:

```
//添加 maven 仓库
repositories {
    maven {
        url 'https://s01.oss.sonatype.org/content/repositories/snapshots/'
    }
}
//添加依赖库
dependencies {
    implementation
    'com.gitee.chinasoft_ohos:datetimepicker:0.0.1-SNAPSHOT'
}
```

添加依赖后直接在 java 代码中导入对应的包即可,注意,使用这个时间选择框的 Java 类需要实现 OnDateSetListener 接口,重写 onDateSet() 方法响应监听事件,在这个事件中处理对话框返回的时间信息。代码如下:

```java
//选择日期框
private void selectDate(){
    DatePickerDialog datePickerDialog;
    //创建 Calendar 对象获取当前系统时间
    Calendar calendar = Calendar.getInstance();
    int year = calendar.get(Calendar.YEAR);
    int month = calendar.get(Calendar.MONTH);
    int day = calendar.get(Calendar.DAY_OF_MONTH);
    //创建 DatePickerDialog 对象,并将当前系统时间设置为初始日期
    datePickerDialog = DatePickerDialog
            .newInstance(this,DailyReportAbilitySlice.this,year,
            month, day,true);
    datePickerDialog.setVibrate(true);
    //设置日期可选范围
    datePickerDialog.setYearRange(1985,2022);
    //设置监听器,监听日期选择框的变化
    datePickerDialog.setOnDateSetListener(this);
    //显示对话框
    datePickerDialog.show();
}
//日期选择器的相应事件
public void onDateSet(DatePickerDialog datePickerDialog, int year, int month,
                                                                int day) {
    //处理日期选择器返回的时间信息
    Calendar calendar = Calendar.getInstance();
    calendar.set(year, month, day);
    symptoms_info = DateUtil.calendarToString(calendar);
    std_symptoms_info.setText(symptoms_info);
    //清除焦点
    std_symptoms_info.clearFocus();
}
```

表单其余属性值的获取不再赘述,读者模仿上述描述的输入框与单选按钮获取输入值的方法实现其他属性值的获取即可。

填写完信息后,单击"提交"按钮,弹出"提交成功!"提示信息,如图 11-10 所示。

图 11-9 日期选择框

图 11-10 提交成功页

11.4.3 数据管理

数据管理主要管理的是用户提交的表单信息,数据保存在 ORM 数据库,实现了对表单信息进行增加和查询的功能,修改和删除表单信息的功能读者可以尝试实现。

1. 数据库以及表的创建

首先创建一个标识数据库的类,并继承 OrmDatabase 类,再通过@Database 注解内的 entities 属性指定这个数据库有哪些实体数据类,对应数据库中数据表。本案例的数据库 StudentDB 类代码如下:

```
@Database(entities = {Student.class}, version = 1)
public class StudentDB extends OrmDatabase {
    public int getVersion(){return 0;}

    public RdbOpenCallback getHelper(){ return null;}
}
```

关于注解的使用在 8.3 节已介绍过,因此不再赘述。

在上述代码中,StudentDB 指定了一个名为 Student 的实体数据类,因此要创建这样一个类,通过注解@Entity 指定该类为一个实体数据类,其私有属性即为数据表的列名,并通过注解@PrimaryKey(autoGenerate = true)设置一个属性为主键且自动增长，Student 类的部分代码如下：

```java
//指明该类是一个实体类,表名为 student
@Entity(tableName = "student")
public class Student extends OrmObject {
    //设置 key 为主键并自动增长
    @PrimaryKey(autoGenerate = true)
    private Integer key;
    //其他属性
    private String id;
    private String name;
    private String gender;
    private String college;
    private String location;
    private Boolean in_hangzhou;
    private String out_hangzhou_reason;
    private String phone_number;
    private double temperature;
    private Boolean symptoms;
    private String symptoms_info;
    private Boolean go_to_high_risk_areas;
    private String hangzhou_health_code;
    private Boolean vaccination;
    private Boolean commitment;
    private String submit_date;
    //添加各个属性的 getter()和 setter()方法
    ...
}
```

所有属性的 getter()和 setter()方法可以利用 IDE 的 Generate 插件自动生成。

2. 操作数据库

操作数据库需要用到对象数据操作接口 OrmContext，在实际的开发中会将对数据库操作的方法封装在一个工具类中，起到复用代码的作用，本案例封装了一个名为 StudentUtil 的类，用于对用户提交的表单数据进行操作，实现添加数据和查询数据的两个公有静态方法。利用 OrmContext 提供的 insert()方法和 query()方法可以很容易地实现添加和查询的操作，而不用写复杂的 SQL 语句。代码如下：

```java
public class StudentUtil {
    //插入用户信息
    public static Boolean insertStudentMsg(Student student,
                                           DatabaseHelper dh){
        Boolean result = false;
        try{
            //获取 OrmContext 对象
            OrmContext ormContext = dh.getOrmContext
```

```
            ("StudentDB", "StudentDB.db", StudentDB.class);
    //开始插入
    if(ormContext.insert(student)) {
        //插入成功
        result = true;
    } else {
        //插入失败
        result = false;
    }
    //持久化数据库
    ormContext.flush();
    ormContext.close();
    }catch (Exception e){
        e.printStackTrace();
    }
    return result;
}
//根据用户 Id 查询用户数据
public static List<Student> queryStudentMsgById(DatabaseHelper dh,
                                                String Id){
    List<Student> students = null;
    try{
        OrmContext ormContext = dh.getOrmContext
            ("StudentDB", "StudentDB.db", StudentDB.class);
        OrmPredicates predicates =
            ormContext.where(Student.class).equalTo("id",Id);
        //按查询条件执行查询
        students = ormContext.query(predicates);
        if(students.isEmpty()) {
            return null;
        }
        ormContext.close();
    }catch (Exception e){
        e.printStackTrace();
    }
    return students;
}
}
```

在 MainAbilitySlice 中添加两个静态的公有方法 insert（Student student）和 query（String Id），供其他 Ability 使用，保证用同一个 DataBaseHelper 实例生成的 OrmContext 对象对数据库进行操作。两个方法的代码如下：

```
public static boolean insert(Student student){
    //直接调用封装好的 insertStudentMsg 方法即可
    return StudentUtil.insertStudentMsg(student, databaseHelper);
}
```

```java
public static List<Student> query(String Id){
    //直接调用封装好的queryStudentMsgById方法即可
    return StudentUtil.queryStudentMsgById(databaseHelper, Id);
}
```

在其他页面只需调用这两个静态方法即可操作 StudentDB。

11.4.4 表单提交

表单提交实际上是将收集到的表单数据插到数据库中，为考虑用户体验，提交成功与否需要给出一个提示信息，如图 11-10 所示。弹出的对话框是一个设置了自定义布局的对话框，且单击对话框以外的区域无法关闭该对话框，只能单击返回主界面的按钮。具体实现如下代码：

```java
//提交用户信息
private void submitStudentInfo(){
    //收集表单数据
    setStudentMsgs();
    //执行插入
    if(MainAbilitySlice.insert(student)){
        //若插入成功则弹出对话框提示
        CommonDialog dialog = new CommonDialog(getContext());
        //动态加载布局文件 Layout_my_toastdialog.xml
        Component container =LayoutScatter.getInstance(getContext())
            .parse(ResourceTable.Layout_my_toastdialog, null, false);
        dialog.setContentCustomComponent(container);
        //设置对话框的尺寸
        dialog.setSize(MATCH_CONTENT, MATCH_CONTENT);
        //关闭单击对话框以外的区域关闭对话框的功能
        dialog.setAutoClosable(false);
        //获取对话框中的按钮
        Button back_mainlayout =(Button)container.findComponentById
            (ResourceTable.Id_back_mainlayout);
        //给按钮添加单击事件返回主界面
        back_mainlayout.setClickedListener(component -> getAbility()
            .terminateAbility());
        //弹出对话框
        dialog.show();
    }
}
```

11.4.5 通行码领取

用户在进入通行码领取界面时，需要从数据库中查询用户填写的每日一报信息，若能查询到该用户当日填报的信息，则界面展示绿码和疫苗接种信息等信息，否则展示红码，并提示用户前往每日一报填写当日的健康状况信息，效果如图 11-11 或图 11-12 所示，实现代码如下：

图 11-11 当日未填写每日一报　　图 11-12 当日已填写每日一报

```java
public class LoginAbilitySlice extends AbilitySlice {
    @Override
    public void onStart(Intent intent) {
        super.onStart(intent);
        super.setUIContent(ResourceTable.Layout_login);
        //组件初始化
        Button login =(Button) findComponentById
            (ResourceTable.Id_login_button);
        TextField name_textview=(TextField) findComponentById
            (ResourceTable.Id_name_textField);
        TextField id_textview=(TextField) findComponentById
            (ResourceTable.Id_id_text_field);
        //设置按钮监听事件
        login.setClickedListener(new Component.ClickedListener() {
            @Override
            public void onClick(Component component) {
                Intent intent = new Intent();
                Operation operation = new Intent.OperationBuilder()
                    .withBundleName(getBundleName())
                    .withAbilityName("MainAbility")
                    .build();
                intent.setOperation(operation);
                //携带参数 getText()获取输入框的值
```

```
                intent.setParam("stdName",name_textview.getText());
                intent.setParam("stdId",id_textview.getText());
                //启动 MainAbility
                startAbility(intent);
            }
        });
    }
}
```

到此,工大通所有功能已经基本实现,但还可以进一步完善,读者可以尝试从以下几个方面进行完善:

(1) 添加用户账户表,管理用户账户信息,使登录功能更加真实。
(2) 添加 HarmonyOS 提供的定位功能,方便用户填写位置信息。
(3) 新增用户填写的表单信息记录页面,用户可以看到自己填写的历史记录。

第 12 章

定点巡检

本章同样会利用前十章所学知识开发一个完整的案例。本章案例是一款具有地图定位和打卡功能的 App,根据设备所在位置判断用户是否到达巡检位置,到达巡检位置后提示用户进行打卡,用户可以查看所有的历史打卡记录。综合了布局、定位、设备相机和振动器等硬件调用、第三方地图组件 SDK 调用等知识,本例的逻辑相对简单,对于初学 HarmonyOS App 开发的人员比较友好,能够很好地利用已掌握的知识开发一个比较完整的案例。

通过阅读本章,读者可以掌握:
- 设备定位。
- 设备相机和振动器调用。
- 第三方地图组件调用。

12.1 功能需求分析

本章案例主要有以下几个功能:
(1) 设备定位及地图位置展示。
(2) 振动提示。
(3) 拍照打卡。
(4) 历史打卡记录。

如图 12-1 所示为定点巡检 App 主要的业务流程。

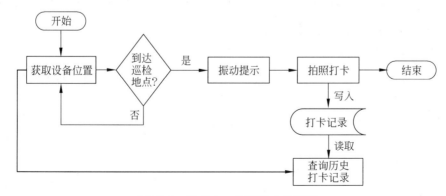

图 12-1 定点 App 主要的业务流程

12.1.1 设备定位及地图位置展示

定点巡检需要实时获取设备所在的位置信息,并与预置的巡检地点位置进行比对来判断设备是否进到巡检范围。为了让用户能够清晰地了解自己所在的位置和目标巡检地点位置,需要在界面上展示地图信息、设备所在位置和巡检地点位置。

12.1.2 振动提示

当用户到达巡检位置时,设备会通过振动提示用户,避免用户错过巡检位置。

12.1.3 拍照打卡

设备振动提示过后,用户可以进行打卡。打卡会调用相机让用户进行拍照作为巡检的记录,同时自动记录打卡时用户的位置和打卡时间。

12.1.4 历史打卡记录

用户可以查看所有的历史巡检打卡记录,记录内容包括巡检时拍摄的照片、巡检地点信息和巡检时间。

12.2 搭建项目框架

12.2.1 项目架构

依照 12.1 节对 App 的主要功能的分析,将整个 App 进行模块划分,具体可以分为以下几个模块:

(1) 定位模块:集成设备实时定位和巡检地点定位功能。

(2) 振动提示模块:发现并调用设备的振动器。

(3) 拍照打卡模块:调用设备的相机模块以支持用户拍照,并将拍摄的照片作为巡检的记录同巡检地点和巡检时间一同上传数据库。

(4) 记录展示模块:查询并展示所有的打卡记录内容。

(5) 数据管理模块:对数据库进行管理,支持初始化及增删改查功能的实现和调用。

(6) 日志管理模块:对系统的日志信息进行集中管理。

如图 12-2 所示为定点巡检的模块架构。

12.2.2 权限设置

定点巡检需要使用设备的定位、相机、地图界面展示等功能,需要获取对应的权限。

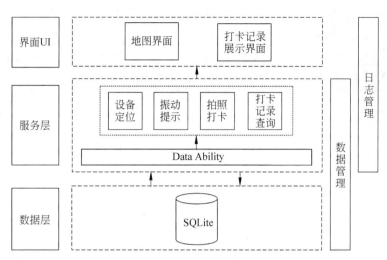

图 12-2 定点巡检模块架构

```
"reqPermissions": [
    {
        "usedScene": {
            "ability": ["com.example.harmonysearchsdk.MainAbility"],
            "when": "always"
        },
        "reason": "request internet",
        "name": "ohos.permission.INTERNET"
    },
    {
        "reason": "usemap",
        "usedScene": {
            "ability": ["com.example.harmonysearchsdk.MainAbility"],
            "when": "always"
        },
        "name": "ohos.permission.GET_BUNDLE_INFO"
    },
    {
        "reason": "",
        "usedScene": {
            "ability": ["com.example.zjutcheckmap.LocationMapAbility"],
            "when": "always"
        },
        "name": "ohos.permission.LOCATION"
    },
    {
        "name": "ohos.permission.VIBRATE",
        "reason": "",
        "usedScene": {
            "ability": ["com.example.zjutcheckmap.LocationMapAbility"],
```

```
            "when": "always"
        }
    },
    {"name": "ohos.permission.WRITE_USER_STORAGE"},
    {"name": "ohos.permission.READ_USER_STORAGE"},
    {"name": "ohos.permission.MICROPHONE"},
    {"name": "ohos.permission.CAMERA"}
]
```

同时在调用 Ability 的过程中也要动态地添加相应的权限。

```java
private void requestPermissions() {
    String[] permissions = {
        SystemPermission.WRITE_USER_STORAGE,
        SystemPermission.READ_USER_STORAGE,
        SystemPermission.CAMERA,
        SystemPermission.MICROPHONE,
        SystemPermission.LOCATION
    };
    requestPermissionsFromUser(Arrays.stream(permissions)
        .filter(permission -> verifySelfPermission(permission) != IBundleManager
        .PERMISSION_GRANTED).toArray(String[]::new), 0);
}
```

12.3 页面设计

12.3.1 地图界面

由于定点巡检 App 需要让用户清晰地了解自己所在的位置和目标巡检地点位置，所以必须要有一个地图界面来展示设备当前所处的位置和目标巡检地点的位置。

地图界面使用了高德地图组件提供的地图 UI，预定的巡检地点需要在载入地图界面时同时设定。定点巡检的地图界面如图 12-3 所示，加载高德地图界面的方法在 12.4.2 节中详细介绍。

12.3.2 打卡拍照界面

当设备进入到巡检地点范围内以后，用户单击"打卡"按钮就会跳转到打卡拍照界面进行打卡操作。由于打卡需要拍照作为记录，所以要调用设备的相机模块，调用相机的方法在 12.4.4 节中详细介绍，界面如图 12-4 所示。

12.3.3 打卡记录界面

用户如需查看历史打卡记录，在地图界面单击"打卡记录"按钮就可以跳转到打卡记录界面，界面会把所有的打卡记录都查询出来并展示给用户，打卡记录界面如图 12-5 所示。

第 12 章 定点巡检

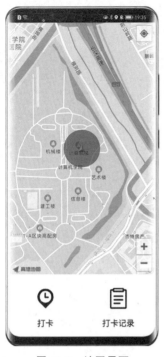

图 12-3 地图界面

图 12-4 打卡拍照界面

图 12-5 打卡记录界面

打卡记录界面的布局使用了 ListContainer 组件，代码如下：

```xml
<?xml version="1.0" encoding="utf-8"?>
<DirectionalLayout
    xmlns:ohos="http://schemas.huawei.com/res/ohos"
    ohos:height="match_parent"
    ohos:width="match_parent"
    ohos:orientation="vertical">
    <Text
        ohos:height="match_content"
        ohos:text_alignment="center"
        ohos:width="match_parent"
        ohos:padding="4vp"
        ohos:text="打卡记录"
        ohos:text_size="30fp"
        ohos:layout_alignment="center"
        ohos:text_color="white"
        ohos:background_element="#1262b3"/>
    <ListContainer
        ohos:id="$+id:list_container"
        ohos:height="match_parent"
        ohos:width="match_parent"
        ohos:layout_alignment="horizontal_center"/>
</DirectionalLayout>
```

ListContainer 组件中的每行填充内容布局如下。

```xml
<?xml version="1.0" encoding="utf-8"?>
<DirectionalLayout
    xmlns:ohos="http://schemas.huawei.com/res/ohos"
    ohos:height="150vp"
    ohos:width="match_parent"
    ohos:orientation="horizontal">
    <Text
        ohos:id="$+id:item_time"
        ohos:height="match_content"
        ohos:text_alignment="center"
        ohos:width="0vp"
        ohos:padding="4vp"
        ohos:text="Item0"
        ohos:multiple_lines="true"
        ohos:text_size="20fp"
        ohos:layout_alignment="center"
        ohos:weight="0.7"/>
    <Image
        ohos:id="$+id:picture"
        ohos:height="80vp"
        ohos:width="0vp"
        ohos:layout_alignment="vertical_center"
        ohos:scale_mode="zoom_center"
        ohos:weight="0.3"/>
</DirectionalLayout>
```

12.4 功能实现

12.4.1 数据管理

实现定点巡检 App 的功能第一步就是做好数据管理。由于定点巡检 App 的功能中包含了打卡记录的写入和读取，使用 DataAbility 做一个数据库的访问封装非常有必要。如果读者对 DataAbility 的内容比较陌生，建议先学习第 9 章的内容。

首先需要定义打卡记录的数据表实体类，RecordDB 代码如下：

```java
@Database(entities = {Record.class}, version = 1)
public class RecordDB extends OrmDatabase {

    @Override
    public int getVersion() {
        return 1;
    }

    @Override
```

```
        public RdbOpenCallback getHelper() {
            return null;
        }
    }
}
```

Record 代码如下:

```
@Entity(tableName = "record")
public class Record extends OrmObject {
@PrimaryKey(autoGenerate = true)
    private Integer id;
    private String time;
    private String location;
    private String latitude;
    private String longitude;
    private String picUrl;
    //省略 get()和 set()方法
}
```

定义好数据库表的实体类,第二步就可以创建 DataAbility 了,案例中创建的是名为 RecordDataAbility 的 DataAbility。创建完成以后,首先要在 onStart()方法中初始化数据库并维持一个数据库连接,代码如下:

```
private static final String DATABASE_NAME = "Record.db";
private static final String DATABASE_NAME_ALIAS = "Record";
private static OrmContext ormContext = null;

public void initDB(Context context) {
    StoreConfig storeConfig = StoreConfig.newDefaultConfig
    ("record.sqlite");
    RdbOpenCallback rdbOpenCallback = new RdbOpenCallback() {
        @Override
        public void onCreate(RdbStore rdbStore) {
            rdbStore.executeSql("CREATE TABLE IF NOT EXISTS record
            (id INTEGER PRIMARY KEY, time VARCHAR(100), location VARCHAR(100),
            latitude VARCHAR(100), picUrl VARCHAR(100),
            longitude VARCHAR(100))");
        }
        @Override
        public void onUpgrade(RdbStore rdbStore, int i, int i1) {
        }
    };
    DatabaseHelper helper = new DatabaseHelper(context);
    RdbStore store = helper.getRdbStore(storeConfig, 1, rdbOpenCallback,
                                        null);
    try {
        RdbPredicates rdbPredicates = new RdbPredicates("record");
```

```
        store.close();
    } catch (Exception e) {}
    ormContext = helper.getOrmContext(DATABASE_NAME_ALIAS,
                            DATABASE_NAME, RecordDB.class);
}

@Override
public void onStart(Intent intent) {
    super.onStart(intent);
    initDB(this);
}
```

完成数据库初始化和数据库连接以后,实现对数据库的增删改查操作,代码如下:

```
@Override
public ResultSet query(Uri uri, String[] columns, DataAbilityPredicates
                                                    predicates) {
    OrmPredicates ormPredicates = DataAbilityUtils.createOrmPredicates
    (predicates, Record.class);
    return ormContext.query(ormPredicates, columns);
}

@Override
public int insert(Uri uri, ValuesBucket value) {
    Record record = new Record();
    record.setId(value.getInteger("id"));
    record.setTime(value.getString("time"));
    record.setLocation(value.getString("location"));
    record.setLatitude(value.getString("latitude"));
    record.setLongitude(value.getString("longitude"));
    record.setPicUrl(value.getString("picUrl"));
    try {
        ormContext.insert(record);
        ormContext.flush();
    } catch (Exception e) {
        return 0;
    }
    return 1;
}

@Override
public int delete(Uri uri, DataAbilityPredicates predicates) {
    OrmPredicates ormPredicates = DataAbilityUtils.createOrmPredicates
    (predicates, Record.class);
    int value = ormContext.delete(ormPredicates);
    ormContext.flush();
    DataAbilityHelper.creator(this, uri).notifyChange(uri);
```

```
        return value;
    }

    @Override
    public int update(Uri uri, ValuesBucket value, DataAbilityPredicates
                                                         predicates) {
        OrmPredicates ormPredicates = DataAbilityUtils.createOrmPredicates
        (predicates, Record.class);
        int result = ormContext.update(ormPredicates, value);
        ormContext.flush();
        DataAbilityHelper.creator(this, uri).notifyChange(uri);

        return result;
    }
```

12.4.2 定位及地图展示

地图模块引入并使用了高德地图的组件进行实现,下面详细说明引入高德地图的步骤。

1. 获取应用的 appId

使用高德地图需要用应用的 appId 来申请高德的 API Key,应用的 appId 需要通过运行以下的代码获取。

```
getApplicationContext().getBundleManager().getBundleInfo(getBundleName(),
                      0).getAppId();
```

2. 申请高德 API Key

使用第一步中获取的 appId 在高德地图官方平台上申请对应的 key。获取到 key 以后把 key 设置给高德地图 SDK,代码如下:

```
String key = ""; //此处是在高德地图官方平台上申请的 API Key
ServiceSettings.getInstance().setApiKey(key);
MapsInitializer.setApiKey(key);
```

3. 在 config.json 中声明需要的权限

代码如下:

```
"reqPermissions": [
    {
        "usedScene": {
            "ability": ["com.example.harmonysearchsdk.MainAbility"],
            "when": "always"
        },
        "reason": "request internet",
        "name": "ohos.permission.INTERNET"
```

```
    },
    {
        "reason": "usemap",
        "usedScene": {
            "ability": ["com.example.harmonysearchsdk.MainAbility"],
            "when": "always"
        },
        "name": "ohos.permission.GET_BUNDLE_INFO"
    }]
```

4. 添加地图开发包依赖

调用高德地图的组件需要导入高德地图提供的依赖 har 包，har 包可以在高德地图的官方平台下载获得（https://lbs.amap.com/api/harmonyos-sdk/download）。

获取到 har 包以后将其放到工程下的 libs 目录中，然后在 build.gradle 文件下加入依赖包。

```
dependencies {
    implementation fileTree(dir: 'libs', include: ['*.jar', '*.har'])
}
```

到此为止已经实现了高德地图的引入，之后就可以调用高德地图的 SDK 来实现定点巡检 App 的地图展示和巡检地点标注功能，使用下面的代码对地图进行初始化，同时对巡检地点进行标注。

```
private void initMapView() {
    mapView = findComponentById(ResourceTable.Id_mapview);
    mapView.onCreate(null);
    mapView.onResume();
    aMap = mapView.getMap();
    aMap.moveCamera(CameraUpdateFactory.zoomTo(17)); //设置初始缩放

    aMap.setOnMapLoadedListener(new AMap.OnMapLoadedListener() {
        @Override
        public void onMapLoaded() {
            setUpMap();
            CoordinateConverter coordinateConverter = new CoordinateConverter
                    (new com.amap.adapter.content.Context(LocationMapAbility.this));
            coordinateConverter.from(CoordinateConverter.CoordType.ALIYUN);
            //标记巡检地点
            for(MarkedPoint markedPoint: LocationMapConstant
                    .markedPointList) {
                coordinateConverter.coord(new LatLng(markedPoint
                        .getLatitude(), markedPoint.getLongitude()));
                LatLng latLng = coordinateConverter.convert();
                aMap.addMarker(new MarkerOptions().position(latLng));
```

```
            }
        }
    });
}
```

巡检地点的定义如下：

```
public class MarkedPoint {
    private String name;
    private double latitude;
    private double longitude;
    //省略 get()和 set()方法
}
```

高德地图 SDK 上已经包含了设备定位的接口，设备定位直接调用相关方法就可以实现，代码如下：

```
//激活定位
@Override
public void activate(LocationSource.OnLocationChangedListener listener) {
    mListener = listener;
    if(hasPermissionGranted(PERM_LOCATION)) {
        if(locator == null) {
            locator = new Locator(this);
        }
        RequestParam requestParam = new RequestParam(RequestParam
            .PRIORITY_FAST_FIRST_FIX, 0, 0);
        locator.startLocating(requestParam, this);
    }
}
//停止定位
@Override
public void deactivate() {
    mListener = null;
    locator.stopLocating(this);
}
//成功定位后回调
@Override
public void onLocationReport(Location location) {
    if(mListener != null && location != null) {
        //定位返回的是 GPS 坐标,需要转换为高德坐标
        CoordinateConverter coordinateConverter = new CoordinateConverter(new
            com.amap.adapter.content.Context(this));
        coordinateConverter.from(CoordinateConverter.CoordType.GPS);
        coordinateConverter.coord(new LatLng(location.getLatitude(),
            location.getLongitude()));

        LatLng latLng = coordinateConverter.convert();
```

```
            location.setLatitude(latLng.latitude);
            location.setLongitude(latLng.longitude);
            mListener.onLocationChanged(location);       //显示系统小蓝点
        }
    }
```

12.4.3 振动器调用

HarmonyOS 给开发者提供了一系列的接口来调用设备的振动器。在调用接口前需要保证设备拥有足够的权限,在 config.json 中配置对应的权限。

```
"reqPermissions": [
    {
        "name": "ohos.permission.VIBRATE",
        "reason": "",
        "usedScene": {
        "ability": ["com.example.zjutcheckmap.LocationMapAbility"],
            "when": "always"
        }
    }
]
```

配置好权限以后就可以调用相关接口实现振动器的调用,代码如下:

```
private void shakeAlert(int point) {
    VibratorAgent vibratorAgent = new VibratorAgent();
    //查询设备的所有振动器
    List<Integer> vibratorList = vibratorAgent.getVibratorIdList();
    if(vibratorList.isEmpty()) {
        return;
    }
    int vibratorId = vibratorList.get(0);
    //查询指定的振动器是否支持指定的振动效果
    VibrationPattern vibrationOnceEffect = VibrationPattern
    .createSingle(3000, 50);
    boolean vibrateSingleResult = vibratorAgent.start(vibratorId,
    vibrationOnceEffect);
}
```

12.4.4 相机调用

和振动器调用一样,HarmonyOS 同样提供了一系列的接口来调用设备的相机,在 config.json 中配置需要的权限。

```
"reqPermissions": [
    {
        "name": "ohos.permission.WRITE_USER_STORAGE"
```

```
    },
    {
        "name": "ohos.permission.READ_USER_STORAGE"
    },
    {
        "name": "ohos.permission.MICROPHONE"
    },
    {
        "name": "ohos.permission.CAMERA"
    },
    {
        "name": "ohos.permission.LOCATION"
    }
]
```

配置好权限以后就可以调用相关接口实现相机的调用,首先初始化相机调用的各种动作,主要实现代码如下:

```
@Override
public void onStart(Intent intent) {
    super.onStart(intent);
    super.setUIContent(ResourceTable.Layout_main_camera_slice);
    //初始化组件
    initComponents();
    initSurface();
}

private void initSurface() {
    getWindow().setTransparent(true);
    DirectionalLayout.LayoutConfig params = new DirectionalLayout
        .LayoutConfig(ComponentContainer.LayoutConfig.MATCH_PARENT,
    ComponentContainer.LayoutConfig.MATCH_PARENT);

    surfaceProvider = new SurfaceProvider(this);
    surfaceProvider.setLayoutConfig(params);
    surfaceProvider.pinToZTop(false);
    if(surfaceProvider.getSurfaceOps().isPresent()) {
        surfaceProvider.getSurfaceOps().get().addCallback(new
                                        SurfaceCallBack());
    }
    surfaceContainer.addComponent(surfaceProvider);
}

private void initComponents() {
    buttonGroupLayout = findComponentById
            (ResourceTable.Id_directionalLayout);
    surfaceContainer = (ComponentContainer) findComponentById
```

```
            (ResourceTable.Id_surface_container);

    Image takePhotoImage = (Image) findComponentById
            (ResourceTable.Id_tack_picture_btn);

    Image exitImage = (Image) findComponentById(ResourceTable.Id_exit);
            Image switchCameraImage = (Image) findComponentById
            (ResourceTable.Id_switch_camera_btn);

    exitImage.setClickedListener(component -> terminateAbility());
    takePhotoImage.setClickedListener(this::takeSingleCapture);
    takePhotoImage.setLongClickedListener(this::takeMultiCapture);
    switchCameraImage.setClickedListener(this::switchCamera);
}
//省略具体实现过程
```

初始化完成后,依据具体需求实现对应的调用方法即可,这里不再一一列举,读者可以参照本教材给出的案例源码进行学习。

12.4.5　打卡操作

用户在巡检地点进行打卡操作,首先调用设备的相机进行拍照,然后把拍摄的照片和巡检地点、时间信息通过调用 RecordDataAbility 的数据库操作方法记录到数据库中,实现代码如下:

```
private void saveImage(ImageReceiver receiver) {
    File saveFile = new File(getExternalFilesDir(null), "IMG_" + System
                    .currentTimeMillis() + ".jpg");

    ohos.media.image.Image image = receiver.readNextImage();
    ohos.media.image.Image.Component component = image.getComponent
                    (ImageFormat.ComponentType.JPEG);

    byte[] bytes = new byte[component.remaining()];
    component.read(bytes);
    try (FileOutputStream output = new FileOutputStream(saveFile)) {
        output.write(bytes);
        output.flush();
        recordInThePoint(saveFile.getPath());
        showTips(this, "打卡成功");
    } catch (IOException e) {}
}

private void recordInThePoint(String picUrl) {
    Uri uri = Uri.parse("dataability:///com.example.zjutcheckmap
                    .RecordDataAbility");
    DataAbilityHelper helper = DataAbilityHelper.creator(this);
```

```
        ValuesBucket valuesBucket = new ValuesBucket();
        valuesBucket.putString("time", DateUtil.currentDate());
        valuesBucket.putString("location", markedPoint.getName());
        valuesBucket.putString("latitude", "" + markedPoint.getLatitude());
        valuesBucket.putString("longitude", "" + markedPoint.getLongitude());
        valuesBucket.putString("picUrl", picUrl);
        try {
            helper.insert(uri, valuesBucket);
        } catch (DataAbilityRemoteException e) {}
    }
```

12.4.6　打卡记录查询

用户查询历史打卡记录,将会调用RecordDataAbility的数据库查询方法查询对应的数据,实现代码如下:

```
private List<Record> queryRecord() {
    Uri uri = Uri.parse("dataability:///com.example.zjutcheckmap
                .RecordDataAbility");
    DataAbilityHelper helper = DataAbilityHelper.creator(this);
    DataAbilityPredicates queryPredicates = new DataAbilityPredicates()
                .greaterThan("id", 0);

    String[] columns = new String[]{"id", "time", "location", "latitude",
                "longitude","picUrl"};

    ResultSet resultSet = null;
    try {
        resultSet = helper.query(uri, columns, queryPredicates);
    } catch (DataAbilityRemoteException e) {}

    List<Record> recordList = new ArrayList<>();
    if(resultSet != null) {
        while(resultSet.goToNextRow()) {
            Record record = new Record();
            record.setId(resultSet.getInt(0));
            record.setTime(resultSet.getString(1));
            record.setLocation(resultSet.getString(2));
            record.setLatitude(resultSet.getString(3));
            record.setLongitude(resultSet.getString(4));
            record.setPicUrl(resultSet.getString(5));
            recordList.add(record);
        }
        resultSet.close();
    }
    return recordList;
}
```

查询出用户的历史打卡记录以后,把数据填充到界面上的 ListContainer 中,实现代码如下:

```java
private void initListContainer() {
    ListContainer listContainer = (ListContainer) findComponentById
            (ResourceTable.Id_list_container);
    List<Record> recordList = queryRecord();

    ItemMomentProvider sampleItemProvider = new ItemMomentProvider
            (recordList, this);
    listContainer.setItemProvider(sampleItemProvider);
    listContainer.setBindStateChangedListener(new
            Component.BindStateChangedListener() {
        @Override
        public void onComponentBoundToWindow(Component component) {
            sampleItemProvider.notifyDataChanged();
        }
        @Override
        public void onComponentUnboundFromWindow(Component component) {
        }
    });
}
```

第 13 章 多媒体播放器

本章利用前十章所学知识完成一个多媒体播放器 App,该播放器支持对本地音乐和视频进行播放,通过获取本地媒体文件信息,实现音视频的播放,综合了各种布局、组件、接口等知识,个别小结需要用户对本书第 7 章有所了解。

通过阅读本章,读者可以掌握:
➢ 使用 ListContainer 组件展示数据。
➢ 封装一个 Player 播放器。
➢ 用多线程对音频和视频的播放进度控制。

13.1 功能需求分析

本章的功能较为明确,首先需要用户授权应用访问本地媒体文件的权限,随后通过调用 HarmonyOS 提供的媒体查询接口,获取到手机存储空间内的媒体文件,包括音频和视频,并将两者展示在页面上。用户通过单击某一个媒体文件跳转到播放页面。业务流程如图 13-1 所示。

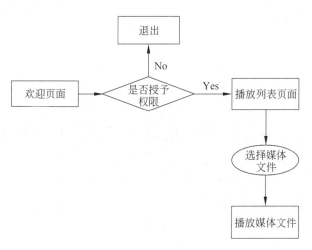

图 13-1 业务流程图

13.1.1 获取读取本地媒体文件权限

应用获取到读取媒体文件的权限是一切的前提,首次进入应用的时候会进入欢迎界面,显示"欢迎!"字样,随后弹出对话框,让用户选择是否授予应用访问本地媒体文件的权限,如果用户选择"始终允许"则进入下一步,否则就直接退出应用。

13.1.2 获取本地媒体文件

通过13.1.1节获取到读取媒体文件的权限后,应用就被授权调用HarmonyOS的相关接口来获取媒体文件信息,在播放页面就可以获取到手机内部存储的媒体文件了,通过列表组件,将获取到的媒体文件信息渲染到页面上。

13.1.3 播放音频和视频

获取到播放列表后,就可以基于获取到的信息做出一些操作,单击播放列表的媒体文件,跳转到播放页面,基于媒体文件信息,可以对播放页面进行渲染,以及初始化封装好的播放器,执行播放任务和媒体控制,包括显示进度条、音量管理、倍速播放、暂停继续。

13.2 搭建项目框架

13.2.1 项目结构

根据13.1节的功能需求分析,本例有以下几个主要的功能模块:
(1) 访问本地媒体文件模块:读取本地媒体文件,获取其信息(包括名称、路径等)。
(2) 文件类型判别模块:用于判别传入的文件属于音频还是视频。
(3) 线程管理模块:用于统一管理执行播放任务的线程。
(4) 播放器模块:用于播放媒体文件,以及媒体控制。

13.2.2 添加应用权限

本例需要获取读写手机存储空间和读取媒体文件的权限,在config.js文件中添加应用权限,代码如下:

```
"reqPermissions": [
    {
        "name": "ohos.permission.WRITE_MEDIA"
    },
    {
        "name": "ohos.permission.WRITE_USER_STORAGE"
    },
    {
        "name": "ohos.permission.READ_USER_STORAGE"
    },
    {
        "name": "ohos.permission.READ_MEDIA"
    },
    {
        "name": "ohos.permission.INTERNET"
    }
]
```

13.2.3 配置相关的 abilities

本例需要用到三个 Ability：WelcomeAbility、MainAbility、PlayerAbility。将它们配置到 config.json 中。

```json
"abilities": [
    {
        "orientation": "unspecified",
        "name": "com.example.mediaplayer.MainAbility",
        "icon": "$media:icon",
        "description": "$string:mainability_description",
        "label": "$string:entry_MainAbility",
        "type": "page",
        "launchType": "standard"
    },
    {
        "skills": [
            {
                "entities": [
                    "entity.system.home"
                ],
                "actions": [
                    "action.system.home"
                ]
            }
        ],
        "orientation": "unspecified",
        "visible": true,
        "name": "com.example.mediaplayer.component.WelcomeAbility",
        "icon": "$media:icon",
        "description": "$string:mainability_description",
        "label": "$string:entry_MainAbility",
        "type": "page",
        "launchType": "standard"
    },
    {
        "orientation": "unspecified",
        "name": "com.example.mediaplayer.component.PlayerAbility",
        "icon": "$media:icon",
        "description": "$string:mainability_description",
        "label": "$string:entry_MainAbility",
        "type": "page",
        "launchType": "standard"
    }
]
```

13.3 界面设计

13.3.1 欢迎界面

欢迎界面主要用于首次打开 App 时弹出对话框,让用户选择是否授予应用访问本地媒体文件的权限。使用 DirectionalLayout 嵌套一个 Text 组件,显示"欢迎!"字样,效果如图 13-2 所示。因此界面布局简单,代码不再赘述。

图 13-2 欢迎界面

13.3.2 媒体列表页

整个页面分为两个部分,一个是选择标签,另一个是媒体文件列表,主要使用 TabList 组件和 ListContainer 组件,将媒体类型分为音乐和视频两类,切换 Tab 就可以查询到不同种类的媒体,并将媒体文件用列表组件展示。使用 DependentLayout 做布局类型,嵌套了标题 Text 和标签切换组件 TabList,以及列表组件 ListContainer。

```
<DependentLayout
    xmlns:ohos="http://schemas.huawei.com/res/ohos"
    ohos:id="$+id:main_ability_ui_root"
    ohos:height="match_parent"
    ohos:width="match_parent">
    <Text
        ohos:id="$+id:text_hello_world"
        ohos:height="match_content"
```

```
        ohos:width="match_content"
        ohos:background_element="$graphic:background_ability_main"
        ohos:layout_alignment="center"
        ohos:horizontal_center="true"
        ohos:text="$string:mainability_HelloWorld"
        ohos:text_size="40vp"
        ohos:visibility="visible"/>
    <TabList
        ohos:normal_text_color="#999999"
        ohos:selected_text_color="#3f3f3f"
        ohos:selected_tab_indicator_color="#FFFFFF"
        ohos:selected_tab_indicator_height="2vp"
        ohos:id="$+id:tab_list"
        ohos:top_margin="10vp"
        ohos:tab_margin="20vp"
        ohos:tab_length="140vp"
        ohos:text_size="20fp"
        ohos:height="36vp"
        ohos:width="match_parent"
        ohos:layout_alignment="center"
        ohos:orientation="horizontal"
        ohos:text_alignment="center"
        ohos:visibility="visible"
        ohos:below="$id:text_hello_world"/>
    <ListContainer
        ohos:top_margin="20vp"
        ohos:height="match_parent"
        ohos:width="match_parent"
        ohos:id="$+id:main_music_list"
        ohos:rebound_effect="true"
        ohos:orientation="vertical"
        ohos:below="$id:tab_list"/>
</DependentLayout>
</DirectionalLayout>
```

ListContainer 组件需要设定好子布局,作为列表中每个项的样式,布局代码如下,效果如图 13-3 所示。

```
<DependentLayout
    xmlns:ohos="http://schemas.huawei.com/res/ohos"
    ohos:height="70vp"
    ohos:width="match_parent"
    ohos:bottom_margin="5vp">
    <Text
        ohos:id="$+id:grey_box"
        ohos:height="match_parent"
        ohos:width="5vp"
```

```xml
        ohos:background_element="$graphic:background_ability_list_item"/>
    <Text
        ohos:id="$+id:item_music_list_id"
        ohos:height="match_parent"
        ohos:width="match_content"
        ohos:background_element="$graphic:background_ability_main"
        ohos:text="1"
        ohos:vertical_center="true"
        ohos:align_parent_left="true"
        ohos:text_size="14vp"
        ohos:left_margin="10fp"/>
    <Text
        ohos:id="$+id:item_music_list_title"
        ohos:height="match_parent"
        ohos:width="match_parent"
        ohos:background_element="$graphic:background_ability_main"
        ohos:text="title"
        ohos:text_size="14vp"
        ohos:end_of="$id:item_music_list_id"
        ohos:left_margin="10fp"/>
    <Text
        ohos:id="$+id:item_music_list_artist_album"
        ohos:height="match_parent"
        ohos:width="match_content"
        ohos:text="artist - album"
        ohos:text_size="14vp"
        ohos:left_margin="200vp"/>
</DependentLayout>
```

图 13-3　播放列表界面

13.3.3 播放器界面

播放器界面是本例的主要功能界面,包含了音视频的播放、标题作者的展示以及音视频的媒体控制等内容。使用 DependentLayout 做界面主布局,内部还嵌套了标题作者、渲染画面、图片、进度条、音量、播放速度、暂停继续等布局,由于布局文件代码较长,此处仅给出部分主要代码,若要详细了解全部布局代码,读者可自行阅读源码。

第一部分是显示媒体文件标题及作者,代码如下:

```
<Text
    ohos:id="$+id:videoTitle_text"
    ohos:text_size="20fp"
    ohos:width="match_content"
    ohos:text="$string:video"
    ohos:layout_alignment="horizontal_center"
    ohos:start_margin="5vp"
    ohos:end_margin="5vp"
    ohos:top_margin="80vp"
    ohos:height="match_content"/>
<Text
    ohos:id="$+id:videoArtist_text"
    ohos:text_size="15fp"
    ohos:width="match_content"
    ohos:text="Anonymous"
    ohos:layout_alignment="horizontal_center"
    ohos:start_margin="5vp"
    ohos:end_margin="5vp"
    ohos:top_margin="10vp"
    ohos:height="match_content"/>
```

第二部分是显示视频渲染组件和默认图片的布局,代码如下:

```
<DirectionalLayout
    ohos:id="$+id:playViewLayout"
    ohos:height="250vp"
    ohos:alignment="center"
    ohos:background_element="#000000"
    ohos:width="match_parent"
    ohos:top_margin="100vp"/>
    <Image
        ohos:id="$+id:album_image"
        ohos:height="250vp"
        ohos:width="250vp"
        ohos:alignment="center"
        ohos:layout_alignment="center"
```

```
        ohos:top_margin="100vp"
        ohos:image_src="$media:album_default"
        ohos:visibility="hide"
        ohos:scale_mode="stretch"/>
```

第三部分是显示媒体文件的进度条,使用 Slider 组件,可实现手动拖曳调节播放进度,代码如下:

```
<Slider
    ohos:id="$+id:playbackProgress_slider"
    ohos:weight="10"
    ohos:height="30vp"
    ohos:width="match_parent"
    ohos:progress_color="red">
</Slider>
```

第四部分是音量管理,同样是使用了 Slider 组件,可以实现手动拖曳调节音量。

```
<Slider
    ohos:id="$+id:playbackVolume_slider"
    ohos:weight="9"
    ohos:height="30vp"
    ohos:width="match_parent">
</Slider>
```

第五部分是播放速度管理,使用了 RadioContainer 及 RadioButton 组件。

```
<RadioContainer
    ohos:id="$+id:playbackSpeed_radioContainer"
    ohos:height="match_content"
    ohos:width="match_content"
    ohos:weight="3"
    ohos:layout_alignment="left"
    ohos:orientation="horizontal"
    ohos:left_margin="4vp">
    <RadioButton
        ohos:id="$+id:playbackSpeed_radioButton1"
        ohos:height="40vp"
        ohos:width="match_content"
        ohos:text="x0.5"
        ohos:text_size="15fp"
        ohos:check_element="$graphic:checkbox_check_element"/>
    //RadioButton组件可复用,此处不再赘述
    ……
</RadioContainer>
```

最终的实现效果如图 13-4 所示。

图 13-4 播放器界面

13.4 功能实现

13.4.1 欢迎界面及权限授予

13.3.1 节讲到,首次进到 App 时会进到欢迎界面,并弹出对话框让用户确认是否授予访问本地媒体文件的权限,这里实际上调用了 HarmonyOS 为开发者提供的 requestPermissionsFromUser()方法,调用该方法会在页面显示对话框。重载父类的 onRequestPermissionsFromUserResult()方法,获取回调的参数,这样就能知道用户选择的结果,如果用户选择"始终允许",则会跳转至媒体列表页面,否则直接退出应用。

```
//判断是否已经授予了权限,如果有则直接跳转,否则要先授予
if(verifySelfPermission(PerMissionUtil.PERMISSION_READER)!=IBundleManager
                        .PERMISSION_GRANTED){
    //PerMissionUtil 是封装权限的类,读者可以自行阅读源码
    if(canRequestPermission(PerMissionUtil.PERMISSION_READER)){
        requestPermissionsFromUser(
            new String[] {PerMissionUtil.PERMISSION_READER,
            PerMissionUtil.PERMISSION_WRITER,
            PerMissionUtil.PERMISSION_READER_MEDIA},1);
    }
}else{
    skipMainAbility();
}
```

```java
@Override
public void onRequestPermissionsFromUserResult(int requestCode,
String[] permissions, int[] grantResults) {
    super.onRequestPermissionsFromUserResult(requestCode, permissions,
                                            grantResults);
    switch (requestCode){
        case 1:
            skipMainAbility();
            break;
        default:
            System.exit(0);
            break;
    }
}
```

13.4.2 媒体列表及获取本地媒体文件

媒体列表界面会显示获取到的媒体文件信息，并以列表的形式展现出来。

获取媒体文件需要先自定义数据结构：MusicBean 类，用于描述每个媒体文件的数据结构，包含 ID、标题、作者、文件路径等内容。

```java
public class MusicBean {
    String id;            //ID
    String title;         //标题
    String artist;        //作者
    String album;         //专辑名称
    String path;          //文件路径
    long duration;        //时长
    //可以用工具自行生成 getter() 和 setter() 方法和构造函数
}
```

为了给 ListContainer 组件添加内容，需要编写 MusicProvider 类继承 BaseItemProvider 类，编写其构造器和重载父类的 getComponent()、getItem()、getCount()、getItemId()等方法。

```java
public class MusicProvider extends BaseItemProvider {
    private static final String TAG = "MusicProvider";
    private List<MusicBean> mMusicBeans;
    private AbilitySlice slice;
    public MusicProvider(List<MusicBean> mMusicBeans, AbilitySlice slice) {
        this.mMusicBeans = mMusicBeans;
        this.slice = slice;
    }
    @Override
    public int getCount() {
        return mMusicBeans == null ? 0: mMusicBeans.size();
    }
```

```java
@Override
public Object getItem(int position) {
    if(mMusicBeans!=null&&position>=0&&position<mMusicBeans.size) {
        return mMusicBeans.get(position);
    }
    return null;
}
@Override
public long getItemId(int position) {
    return position;
}
@Override
public Component getComponent(int position, Component component,
ComponentContainer componentContainer) {
    if(getCount() <= 0) return null;
    if(component == null) {
        component = LayoutScatter.getInstance(slice).parse
            (ResourceTable.Layout_item_music_list,null,false);
    }
    Text tx_number, tx_title,tx_artist_album;
    MusicBean bean = mMusicBeans.get(position);
    tx_number=(Text)component.findComponentById
            (ResourceTable.Id_item_music_list_id);
    tx_title=(Text)component.findComponentById
            (ResourceTable.Id_item_music_list_title);
    tx_artist_album=(Text)component.findComponentById
            (ResourceTable.Id_item_music_list_artist_album);
    LogUtil.debug(TAG,bean.toString());
    tx_number.setText(bean.getId());
    tx_title.setText(bean.getTitle());
    tx_artist_album.setText(bean.getArtist());
    return component;
}
}
```

为 ListContainer 设置 MusicProvider 对象,用于给列表添加内容。

```java
private void UpdateProvider(List<MusicBean> beans) {
    if(beans == null || mMusicList == null) return;
    mMusicprovider = new MusicProvider(beans,this);
    mMusicList.setItemProvider(mMusicprovider);
}
```

在本例中,通过切换 TabList 可以选择获取音频和视频,本质是利用 TabList 的下标来判断当前需要获取的是音频还是视频。

定义一个 SongSheet 接口,用来表示获取媒体资源。

```java
public interface SongSheet {
    void showSheetMusic(OnLocalSheetListener onLocalSheetListener,
    Context context, int position);
    public interface OnLocalSheetListener{
        void complete(List<MusicBean> beans);
    }
}
```

定义一个实现类 AllSongSheetModel，来实现 SongSheet 接口，重载 showSheetMusic() 方法，以及实现 getSheetMusic() 方法和 getSheetVideo() 方法来获取本地媒体文件列表，并将获取到的媒体列表传入上层定义的 OnLocalSheetListener 接口，这个接口的 complete() 方法由 MainAbilitySlice 去实现。

```java
public class AllSongSheetModel implements SongSheet{
    private static final String TAG = "AllSongSheetModel";
    private List<MusicBean> musicBeanList;
    @Override
    public void showSheetMusic(OnLocalSheetListener onLocalSheetListener,
    Context context, int position) {
        if(position == 0){
            onLocalSheetListener.complete(getSheetMusic(context))
        }else{
            onLocalSheetListener.complete(getSheetVideo(context))
        }
    }
    private List<MusicBean> getSheetMusic(Context context){
        List<MusicBean> mDates = new ArrayList<>();
        if(context == null) return mDates;
        context = context.getApplicationContext();
        DataAbilityHelper helper = DataAbilityHelper.creator(context);
        Uri uri = AVStorage.Audio.Media.EXTERNAL_DATA_ABILITY_URI;
        try{
            ResultSet query = helper.query(uri,null,null);
            if(query == null) return mDates;
            int i = 1;
            while(query.goToNextRow()){
                String duration = query.getString(query.getColumnIndexForName
                        (AVStorage.Audio.Media.DURATION));
                int index = query.getInt(query.getColumnIndexForName
                        (AVStorage.Audio.Media.ID));
                String title = query.getString(query.getColumnIndexForName
                        (AVStorage.Audio.Media.TITLE));
                String artist=query.getString(query.getColumnIndexForName
                        ("artist"));
                String album=query.getString(query.getColumnIndexForName
                        ("album"));
```

```
                String path = query.getString(query.getColumnIndexForName
                         (AVStorage.Audio.Media.DATA));
                MusicBean bean = new MusicBean(String.valueOf(i++),title,
                         artist,album,path,path,index,Long.parseLong(duration));
                mDates.add(bean);
             }
        } catch (DataAbilityRemoteException e) {
            e.printStackTrace();
        }
        return mDates;
    }
    private List<MusicBean> getSheetVideo(Context context) {
        //获取视频列表的方法与音乐类似只需将 Uri 改为 EXTERNAL_DATA_ABILITY_URI
        Uri uri = AVStorage.Video.Media.EXTERNAL_DATA_ABILITY_URI;
    }
```

在 MainAbilitySlice 中的生命周期函数 onActive() 中实现 complete() 方法，也就是将获取到的数据渲染到 ListContainer 中。

```
public void onActive() {
    super.onActive();
    if(mSongSheet != null){
        mSongSheet.showSheetMusic(new SongSheet.OnLocalSheetListener() {
            @Override
            public void complete(List<MusicBean> beans) {
                UpdateProvider(beans);//上文提到了 UpdateProvider 函数
            }
        //currentPosition 用于标记当前要获取的是音乐还是视频
        },getApplicationContext(), currentPosition);
    }
}
```

设置 ListContainer 组件的单击响应事件，带参数(包含标题、时长、作者、相册、路径)跳转到播放器页面。

```
mMusicList.setItemClickedListener((container, component, position, id)->{
    //设置 ListContainer 单个 item 的单击事件：传参跳转到 PlayerAbility
    MusicBean bean = (MusicBean) mMusicList.getItemProvider()
            .getItem(position);
    Intent intent = new Intent();
    Operation operation = new Intent.OperationBuilder().withDeviceId("")
            .withBundleName("com.example.mediaplayer")
            .withAbilityName("component.PlayerAbility").build();
    intent.setParam("title", bean.getTitle());
    intent.setParam("duration", String.valueOf(bean.getDuration()));
    intent.setParam("artist", bean.getArtist());
    intent.setParam("album", bean.getAlbum());
```

```
            intent.setParam("path", bean.getPath());
            intent.setOperation(operation);
            startAbility(intent);
        });
```

13.4.3 封装一个 Player 播放器类

播放器组件 VideoPlayerPlugin 是该项目的核心组件,主要实现了音频和视频的播放、暂停、恢复、倍速播放、音量管理、跳转播放的功能。

播放器组件需要引入线程管理功能。在本例中封装了 ThreadPoolManager 类用于管理线程。

```java
public class ThreadPoolManager {
    private static final int CPU_COUNT = 20;
    private static final int CORE_POOL_SIZE = CPU_COUNT + 1;
    private static final int MAXIMUM_POOL_SIZE = CPU_COUNT * 2 + 1;
    private static final int KEEP_ALIVE = 1;
    private static ThreadPoolManager instance;
    private ThreadPoolExecutor executor;
    private ThreadPoolManager() {
        if(executor == null) {
            executor = new ThreadPoolExecutor(CORE_POOL_SIZE,
            MAXIMUM_POOL_SIZE, KEEP_ALIVE, TimeUnit.SECONDS,
            new ArrayBlockingQueue<>(20), Executors.defaultThreadFactory(),
            new ThreadPoolExecutor.AbortPolicy());
        }
    }
    public static synchronized ThreadPoolManager getInstance() {
        if(instance == null) {
            synchronized (ThreadPoolManager.class) {
                if(instance == null) {
                    instance = new ThreadPoolManager();
                }
            }
        }
        return instance;
    }
    public void execute(Runnable runnable) {
        executor.execute(runnable);
    }
    public void cancel(Runnable runnable) {
        if(runnable != null) {
            executor.getQueue().remove(runnable);
        }
    }
}
```

首先讲解基础的播放功能,先要判断当前播放器的状态,如果当前播放器正在运行,需要先将其初始化,并设置回调类,然后将播放线程提交到线程池,主要实现代码如下:

```java
public synchronized void play(MusicBean bean, Surface surface) {
    if(videoPlayer != null) {
        videoPlayer.stop();
        videoPlayer.release();
        videoPlayer = null;
    }
    if(videoRunnable != null) {
        ThreadPoolManager.getInstance().cancel(videoRunnable);
    }
    videoPlayer = new Player(context);        //初始化播放器
    setPlayerCallback();                      //设置播放器回调类
    videoRunnable = () -> playInner(bean, surface);
    //将线程提交到线程池
    ThreadPoolManager.getInstance().execute(videoRunnable);
}
private void playInner(MusicBean bean, Surface surface) {
    //获取播放源
    Source source = new Source(bean.getPath());
    videoPlayer.setSource(source);
    videoPlayer.setVideoSurface(surface);
    LogUtil.info(TAG, source.getUri());
    videoPlayer.prepare();
    videoPlayer.play();
    //设置当前状态为正在播放
    playbackState = PlaybackState.VIDEO_PLAYER_PLAYING;
    startTimer();                             //开启定时器,用于同步播放进度
}
```

暂停功能较为简单,就是调用 Player 类的接口,主要代码如下:

```java
public synchronized void pause() {
    if(videoPlayer == null) return;
    videoPlayer.pause();
    //设置视频状态为暂停
    playbackState = PlaybackState.VIDEO_PLAYER_PAUSING;
    endTimer();        //关闭定时器,用于同步播放进度
}
```

恢复播放,跳转播放进度,设置音量等功能与上述流程大致相似,这里不再赘述,读者可自行参照源码。

接下来讲解一下进度条的实现,需要在封装的播放器中动态维护一个定时器 timer,当开始播放的时候就调用 startTimer()方法,定时调用更新进度条函数 playbackProgressUpdate()。

```
private void startTimer() {
    if(timer == null) {
        timer = new Timer("PlaybackPorgressTimer");
        timer.schedule(new TimerTask() {
            @Override
            public void run() {
                playbackProgressUpdate();
            }
        }, 200L, 200L);
    }
}
private void endTimer() {
    if(timer == null) return;
    timer.cancel();
    timer = null;
}
private void playbackProgressUpdate() {
    if(eventHandler == null || videoPlayer == null) return;
    int currentPlaybackProgress = videoPlayer.getCurrentTime();
    EventHandler handler = new EventHandler(EventRunner
                            .getMainEventRunner());
    handler.postTask(new Runnable() {
        @Override
        public void run() {
            eventHandler.onPlaybackProgressUpdate(currentPlaybackProgress);
        }
    });
}
```

并且在外部调用的地方实现 onPlaybackProgressUpdate() 函数，就可以将播放器内部的时间数据通过实现接口函数传给外部进度条，实现进度条的状态更新。

```
@Override
public void onPlaybackProgressUpdate(int millisecond) {
    playbackProgressSlider.setProgressValue(millisecond);
    String strCurrentDuration = TimeUtil.stringForTime(millisecond);
    currentDurationText.setText(strCurrentDuration);
}
```

13.4.4 实现 PlayerAbility

PlayerAbility 负责维护播放页面的生命周期，首先在 onStart() 函数内将 MainAbilitySlice 传进来的参数接收到，并保存在一个 MusicBean 对象中，接下来执行一系列初始化函数。

```
public void onStart(Intent intent) {
    super.onStart(intent);
    super.setUIContent(ResourceTable.Layout_player_ability);
    try{
```

```
            if(intent != null){                //接收来自播放列表的参数
                bean = new MusicBean();
                bean.setTitle(intent.getStringParam("title"));
                bean.setArtist(intent.getStringParam("artist"));
                bean.setAlbum(intent.getStringParam("album"));
                bean.setPath(intent.getStringParam("path"));
                bean.setDuration(Long.parseLong(intent
                        .getStringParam("duration")));
                LogUtil.debug(TAG,bean.toString());
            }
            addSurfaceProvider();               //添加渲染组件 SurfaceProvider
            initPlayer();                       //初始化播放器
            initComponents();                   //初始化组件,设置响应事件
            playViewDisplay();                  //判断目前播放的是视频还是音乐
        }catch (Exception e) {
            LogUtil.error(TAG,e.toString());
        }
    }
```

addSurfaceProvider()方法用来渲染视频组件,在第 7 章介绍过 SurfaceProvider 组件,主要用来承载视频的画面,初始化 SurfaceProvider 并且设置回调类。

```
private void addSurfaceProvider() {
    surfaceProvider = new SurfaceProvider(this);
    if(surfaceProvider.getSurfaceOps().isPresent()) {
        surfaceProvider.getSurfaceOps().get().addCallback
                (new SurfaceCallBack());
        surfaceProvider.pinToZTop(true);
    }
}
class SurfaceCallBack implements SurfaceOps.Callback {
    @Override
    public void surfaceCreated(SurfaceOps callbackSurfaceOps) {
        if(surfaceProvider.getSurfaceOps().isPresent()) {
            surface = surfaceProvider.getSurfaceOps().get().getSurface();
            if(isAudio == false){
                play();
            }
        }
    }
}
```

initPlayer()用来初始化播放器,将上下文信息 context 传给播放器组件,以及设置播放器的回调类。

```
private void initPlayer() {
    videoPlayerPlugin = new VideoPlayerPlugin(getApplicationContext());
    setVideoPlayerEventHandler();
}
```

```java
private void setVideoPlayerEventHandler() {
    videoPlayerPlugin.setVideoPlayerEventHandler(new VideoPlayerPlugin
                    .VideoPlayerEventHandler() {
        @Override
        public void onPlaybackProgressUpdate(int millisecond) {
            playbackProgressSlider.setProgressValue(millisecond);
            String strCurrentDuration = TimeUtil.stringForTime(millisecond);
            currentDurationText.setText(strCurrentDuration);
        }
    });
}
```

initComponents()函数用于初始化页面的组件,以及设置其响应函数,与前文介绍类似,读者可自行阅读源码。

最后需要通过playViewDisplay()函数来判断传入的媒体文件属于什么类型,如果是音频文件则需要将playViewLayout组件隐藏,并显示默认的专辑图片组件,否则将专辑图片组件隐藏,显示用于渲染视频的playViewLayout组件。

```java
private void playViewDisplay(){
    String[] res = bean.getPath().split("/");
    //FileTypeUtil.getMimeType方法用于判断文件类型
    String fileType = FileTypeUtil.getMimeType(res[res.length-1]);
    if(fileType.split("/")[0].equals("audio")){
        isAudio = true;
        playViewLayout.setVisibility(Component.HIDE);
        album_default.setVisibility(Component.VISIBLE);
        play();
    }else{
        isAudio = false;
        playViewLayout.setVisibility(Component.VISIBLE);
        album_default.setVisibility(Component.HIDE);
    }
}
```

通过调用play()函数即可开始音视频的播放。注意,当传入的文件类型是视频时,play()函数要等到创建完SurfaceProvider之后才可以调用,也就是在SurfaceProvider的回调函数中进行调用,否则会出现有声音没画面的情况。

```java
class SurfaceCallBack implements SurfaceOps.Callback {
    @Override
    public void surfaceCreated(SurfaceOps callbackSurfaceOps) {
        if(surfaceProvider.getSurfaceOps().isPresent()) {
            surface = surfaceProvider.getSurfaceOps().get().getSurface();
            if(isAudio == false){
                play();
            }
        }
    }
}
```

内容简介

本书系统全面地讲解在鸿蒙操作系统（HarmonyOS）下基于Java的应用程序开发的基础理论知识，通过丰富、详细的案例向读者呈现HarmonyOS应用程序的开发流程。全书共13章。第1章对HarmonyOS的概念、技术特性以及技术架构进行了综合介绍；第2章以一个简单的Hello World工程为例，介绍HarmonyOS应用程序的开发环境、开发工具以及应用的调试过程，并对HarmonyOS的工程结构进行讲解，使读者能更好地切入和理解后续章节的学习内容；第3章详细介绍HarmonyOS应用程序的一大核心——Page Ability，其是完成后续章节学习的基础；第4~6章分别对布局、组件以及对话框进行系统介绍；第7章介绍HarmonyOS应用程序中多媒体的开发过程；第8、9章介绍HarmonyOS应用程序中数据管理和文件管理的部分；第10章介绍HarmonyOS应用程序中后台任务如何通过Service Ability运行；第11~13章分别介绍三个完整的案例（工大通、定点巡检、多媒体播放器），不仅涉及基本的布局、组件、数据管理等基础知识，还涉及了对设备硬件调用等进阶知识，读者可以在这三个案例的基础上进行二次开发，使其功能更加丰富，更具有实用性和应用性。

本书主要面向鸿蒙应用的入门开发人员，也可作为高校教材或培训机构的参考用书。

课件下载·样书申请　清华大学出版社

作业系统
二维码

书　圈

官方微信号

ISBN 978-7-302-63340-2

定价：59.00元